THE CITY

PARENTS AND CHILDREN IN THE
INNER CITY

SOCIOLOGY OF THE CITY

PARENTS AND CHILDREN IN THE INNER CITY

HARRIETT WILSON & G.W. HERBERT,

LONDON AND NEW YORK

First published in 1978

This edition published in 2007
Routledge
2 Park Square, Milton Park, Abingdon, Oxfordshire OX14 4RN
711 Third Avenue, New York, NY 10017

First issued in paperback 2014

Routledge is an imprint of Taylor & Francis Group, an informa business

© 1978 Harriett Charlotte Wilson & Geoffrey William Herbert

All rights reserved. No part of this book may be reprinted or reproduced or utilized in any form or by any electronic, mechanical, or other means, now known or hereafter invented, including photocopying and recording, or in any information storage or retrieval system, without permission in writing from the publishers.

The publishers have made every effort to contact authors and copyright holders of the works reprinted in the *The City* series. This has not been possible in every case, however, and we would welcome correspondence from those individuals or organisations we have been unable to trace.

These reprints are taken from original copies of each book. In many cases the condition of these originals is not perfect. The publisher has gone to great lengths to ensure the quality of these reprints, but wishes to point out that certain characteristics of the original copies will, of necessity, be apparent in reprints thereof.

British Library Cataloguing in Publication Data
A CIP catalogue record for this book
is available from the British Library

Parents and Children in the Inner City
ISBN10: 0-415-41780-5 (volume)
ISBN10: 0-415-41931-X (subset)
ISBN10: 0-415-41318-4 (set)

ISBN13: 978-0-415-41780-8 (volume)
ISBN13: 978-0-415-41931-4 (subset)
ISBN13: 978-0-415-41318-3 (set)

ISBN 13: 978-1-138-87392-6 (pbk)

Routledge Library Editions: The City

PARENTS AND CHILDREN IN THE INNER CITY

HARRIETT WILSON and G.W. HERBERT
with a contribution on incomes
by John Veit Wilson

ROUTLEDGE DIRECT EDITIONS

ROUTLEDGE & KEGAN PAUL
London, Henley and Boston

First published in 1978
by Routledge & Kegan Paul Ltd
39 Store Street,
London WC1E 7DD,
Broadway House,
Newtown Road,
Henley-on-Thames,
Oxon RG9 1EN and
9 Park Street,
Boston, Mass. 02108, USA
Printed in Great Britain by
Thomson Litho Ltd,
East Kilbride, Scotland
© Harriett Charlotte Wilson & Geoffrey William Herbert 1978
No part of this book may be reproduced in
any form without permission from the
publisher, except for the quotation of brief
passages in criticism

British Library Cataloguing in Publication Data

Wilson, Harriett
   Parents and children in the inner city.
   1. Poor - England - Midlands 2. Children
   3. Cost and standard of living - England -
   Midlands 4. Urban economics
   I. Title II. Herbert, G W
   362.7   HD7024   77-30436

   ISBN 0-7100-8715-2

CONTENTS

PROLOGUE	vii
ACKNOWLEDGMENTS	ix

Part one DATA

1	THE PROBLEM DEFINED	3
2	THE FAMILIES AND THE ENVIRONMENT	15
3	THE CONTROL GROUPS AND THE RANDOM SAMPLE	33
4	THE BOYS AT SCHOOL	40
5	THE PLAY-GROUPS	64
6	DELINQUENCY	76

Part two CONVERSATIONS WITH PARENTS

7	SHARED EXPERIENCE AND THE SOCIALIZATION OF CHILDREN	87
8	THE WORLD OF THE CHILD	105
9	ACHIEVEMENT, SELF-RELIANCE AND RESPONSIBILITY	123
10	TOLERANCE AND DISCIPLINE	147

Part three INTERPRETATION

11	CHILD-CENTREDNESS	163
12	THE POVERTY SYNDROME AND CHILD DEVELOPMENT	180

APPENDICES 199

REFERENCES 233

NAME INDEX 241

SUBJECT INDEX 244

PROLOGUE

The parents about whom we have written this book live in the poorest areas of a large city. They are widely dispersed; they do not know one another. There are certain features about their lives that bind them together and make them speak as if they had exchanged their views. They have shared experiences. They are poor. They feel deprived in relation to people they meet in the shops and they see on television. They carry memories of deprived childhoods and parental accounts of poverty. Some trace their family histories back to Ireland, to Wales, to rural labouring; others have been city dwellers for generations and often out of work. 'Those days we were that poor, drinking out of jam jars, having scratchings for dinner. I see my Mum lying on a bed, dead, she was sucking a sugar stick. We all went to Barnardo's except the one that died.' Many come from large families and know the sorrows of premature death, disablement, stillbirth and unwanted pregnancy. They know their limitations and have adapted to failure by lowered expectation. They largely see themselves as the outsider sees them: as rough people, humble, unsuited for positions that hold responsibility. The positions they hold are at the receiving end: they take the jobs that are available, the benefits they are given, the accommodation they are allocated; they accept discretionary decisions about additions to or deductions from benefit, rent allowances, rate rebates, applications for free school meals. They suffer eviction, parting from children, redundancies, the wage stop. They accept and make do. This account of fifty-six families is an attempt to explore the interrelationship between the parents' circumstances and the difficulties encountered by their children.

ACKNOWLEDGMENTS

We should like to express our thanks to those who made this book possible. The initial planning was undertaken under the guidance of Professor B. Cullingworth, at the time Director of the Centre for Urban and Regional Studies, University of Birmingham. The fieldwork was located at the Centre for Child Study, whose Director, C.J. Phillips, we thank for administrative support, for general guidance and contributions to the design and statistical analysis of the psychological work, and for the use of comparison data from his researches in the locality. Much of the analysis was performed on the University of Birmingham computer, and we are grateful to Dr R.D. Lambourne, Professor P.M. Levy, and R. Phillips for computer programmes and advice. The final analysis of the interview data was undertaken at the Centre for Urban and Regional Studies, and we thank Professor Gordon Cherry, at the time Acting Director, for accommodation and administrative assistance.

The Chief Officers of many statutory agencies of the Midlands city in which the work took place played an active part in facilitating access to information, and we thank especially the Director of the Social Services Department, the Chief Education Officer, the Chief Constable, the Principal Probation Officer, the Regional Controller of the Supplementary Benefit Commission, the Principal School Medical Officer, and the many members of their staffs who were involved in the fieldwork.

There are many personal contributions to this book. We should like to thank Dr Mary Brennan for medical examinations of the school boys; Dr David Eversley for critical comments on the presentation of the material; Annette Holman who conducted most of the interviews; Anne-Marie Homer who was our research assistant during much of the work; Jan Rees for voluntary help throughout the field work; John Veit Wilson for an analysis of incomes; and the husband of HW for statistical guidance.

We are indebted to Professor John Newson and Dr Elizabeth Newson for permission to adapt their interview schedules and to use unpublished material for comparative purposes; to Dr D.J. West and Dr H.B. Gibson for permission to use the Conduct Scale and its scoring system; to Professor E.A. Lunzer for the Piagetian tests;

to Beryl Smith and C.J. Phillips for the use of other cognitive tests developed at the Centre for Child Study.

We owe special thanks to the Family Service Unit for their generous advice and guidance at the outset of our fieldwork.

We are deeply grateful for the time patiently given to us by the fifty-six families who enabled us to gain some insight into their lives.

The study was financed by the Home Office, the Department of Health and Social Security, and the Bernard Van Leer Foundation.

Part one

DATA

Chapter 1

THE PROBLEM DEFINED

This book describes a study of families who are known to the social services department of a large Midlands city. The main objective of the study was to explore the development of two children in each family and to relate this to the social and economic circumstances of the home. Each family had one child aged between three and five, who participated in a nursery play group, and also a boy of school age, the main focus of the psychological observations. All families had five or more children, and lived in deprived inner areas of the city. Immigrant families (except Irish) were not included, to keep cultural factors as constant as possible. The families thus formed a sample of people under stress who at some stage had been unable to provide for the care of their children in the way that parents normally do, and this had led to self-referral or referral by some other person or agency to the social services department. Under Section I of the Children's and Young Persons' Act, 1963, the department provides advice and assistance so as to diminish the need to receive children into care. The period of time during which families had contact with the department varied; some of them had received intensive support over a lengthy period of time, others had had short contact amounting to no more than four to six visits, no further action being considered necessary

LARGE FAMILIES

Large families are no longer typical of families in general. Average family size, as a measure of the number of children born to each married woman, has declined from approximately six children in the 1870s to a little over two children in recent years. There is a social-class trend, middle-class families tending to be smaller than working-class families on average, although fertility among professional families almost reached the mean family size for unskilled workers towards the end of the 1960s. In a national sample of children born in 1946 the percentage of children who come from large families (five or more) was found to be 16.5 (Douglas, 1964), and in a sample of children in Nottingham born about 1960 it was 19.0 (Newson and Newson, 1968). In another national sample of children born in 1958 more than one in every six children were found to belong to a family with five or more children (Pringle, Butler and Davie, 1966). The total number of children who grow up in large families is likely to be above two and a half million.

Large families are more typical of the manual working class. Only 6 per cent of families in classes I and II (professional and managerial) had five or more children in the Nottingham sample, in contrast with class IV (semi-skilled manual) with 20 per cent and class V (unskilled manual) with 38 per cent. In a sample of boys from a number of schools in a working-class area of London the percentages of boys with five or more siblings in classes IV and V were found to be 20.0 and 31.1 respectively (West, 1969). This study drew attention to the fact that a working-class community will contain a significantly larger proportion of children from large families compared with the national average: there was a deficit of small families, and almost one quarter of all families had five or more children. Similar findings are reported in four educational priority areas: the percentage of large families was 24 in Denaby, 30 in Deptford, 37 in Liverpool, and 55 in Birmingham (Halsey, 1972). Our sample was thus not untypical of a substantial minority of families living in deprived areas.

CONTACT WITH A SOCIAL SERVICES DEPARTMENT

How many families are in contact with their local authority over problems concerning their children's welfare? The annual report of the Department of Health and Social Security, 1972, states that under Section I a total of 203,863 applications for help or referrals had been made to departments in England and Wales during the preceding year. This number is some 50 per cent above that recorded by the Home Office Children's Department (as it then was) in 1969, when the total was 136,981 applications; and this again constitutes a 30 per cent increase over the figure of 109,245 recorded in 1967. This steadily increasing rate of applications suggests that, in all, over one million families applied for advice or help during the five-year period 1968-72, but it is not an accurate estimate as some authorities do not supply this information, and there may be some duplications of families. A clearer picture of the extent to which families contact the department is given in a study of disadvantaged children who are part of the National Child Development Study of 11,000 children born in 1958 (Wedge and Prosser, 1973). The definition of 'disadvantage' was as follows: the children were in either a one-parent or large family, they were badly housed, and they were in a low-income family. Ten per cent of this disadvantaged group were known to a social services department. In contrast, one in 300 of the remaining 'ordinary' families had contact with the social services.

From this information it appears that only a relatively small sector of families turns to a social services department for help, and they are almost entirely families under great material stress. An indication of the social and economic factors which most often give rise to contact with the department is provided by Packman (1968) who, in a survey of children in care, found that 'illegitimate children, unemployed fathers, the lowest social classes, poorly housed and overcrowded families, newcomers and foreigners were all prominent'. The main factors rendering a family vulnerable appear to be poverty, inadequate housing, and isolation. But what factor or factors cause poverty, drive families with children into poor housing, create isolation and dependence on public services? Packman singles out unemployment and non-skilled occupations as well as the status of immigrants. Schaffer and Schaffer (1968) suggest 'inadequacy' in describing these families as 'inadequate parents bringing up their children inadequately to become inadequate parents in their turn'. The inadequacy is seen as 'a failure to act in a protective capacity'. The term is given a double meaning: it refers to inadequate functioning and to inadequacy of personality. While the former refers to an inability to live up to expected performance, the latter shifts the conceptual frame of reference to one of parental pathology. An inability to conform to expected standards may or may not be due to personality pathology; it may be due to low occupational status and low earning capacity, low per capita income, prolonged unemployment, sickness, or absence of a male wage earner, or other material stress factor. The differentiation between inadequacy of conditions leading to inadequate functioning and inadequacy of personality - however defined - is rarely undertaken. Instead, the complex process of interaction between circumstances and personality is summed up in terms of a causal relationship which postulates the personality as being the primary agent in the entire process, instead of according material stress at least a similar position. Prolonged stress affects mental states and behaviour, but this is rarely seen as an explanation of malfunction; nor is the more complex relationship of subnormality and material stress - which frequently go together - understood as a challenge to inadequate provision.

The complexity of the problem resides not only in the links between material environment and human personality, but also in the making of judgments that have to be made in a study that concerns itself with the functional adequacy of human groups. A family whose main income earner has a low earning capacity may yet function adequately in conditions of full employment, at a wage level above the poverty line, in housing fit for family life, and supported by adequate health and social services. If, on the other hand, low earning capacity leads to frequent unemployment, and social security benefits are pegged at a level beneath the poverty line, such a family may learn to adapt to its stringent circumstances, but some essential element of that adaptation may be inimical to child development. The situations that motivate children in their own homes are governed by the constraints of their material circumstances, and they are different in nature from the situations encountered at school. Children adapt to family and neighbourhood, but in doing so they become maladapted to school. This is not necessarily maladjustment in the clinical sense, although occasionally they may react to maladaptation with behaviour looking like maladjustment - aggression, aimless truancy, withdrawal, passive conformity, and so on - and this may result in referral to a Child Guidance Clinic which is expected to 'do something' about it, as if it were a problem of personality or parent-child relationships. An alternative to this assumption of maladjustment is the view that there is a 'pathology of a disorganised situation' (Boxall, 1973), i.e., a deficiency in the environment, which must be remedied by compensating measures in an educational setting. A better understanding of the home situation is essential so that the problems presented by these children at school can be more clearly understood.

SOCIOLOGICAL ASSUMPTIONS

The search for causative factors of social deprivation is governed by basic assumptions that are rarely made explicit. They have to do with the organization of society and the role of individuals in relation to the community. One such assumption is that we live in a society with largely shared values that are functionally related to its economic and social structure (Parsons, 1961). These values are largely concerned with the problems of social order. On the level of the individual, character traits that are appreciated include ambition, the will to work hard, and the will to defer pleasures, to do one's duty and to learn to carry responsibility. This ethic is suspicious of change, emphasizes conformity and stresses personality development. Adherents of this viewpoint place much importance on the socialization process of early childhood which facilitates the internalization of these values; maturational processes absorb these values so that they become part of the adult personality. The societal system is supported by an intricate mechanism of social controls, which range from sanctions at the level of individual relationships to enforcement of behavioural standards by the law. The poor, the socially deprived, the disadvantaged sectors of the community are seen as people who lack the ability to function as independent self-supporting members of society. Explanations of their state vary, but three main groups are discernible. There are, first, interpretations in terms of a pathology. This school of thought tends to stress genetic aspects of the personality and emphasizes subnormality, mental illness, and generational aspects of their status (see, for instance, Tonge, James and Hillam, 1975). A second school of thought stresses cultural aspects of personality development and explains low status in terms of 'cultural deprivation', a deficit that needs to be compensated for, a gap that needs to be filled by educational enrichment (see, for instance, Ferguson and Williams, 1969). A third, less coherent, variety of thought is expressed by moralists who see persons of low status not as deprived but as depraved, the feckless, the workshy, the criminal, who are in need of stricter discipline and more effective societal systems of control. This view is aptly described by Valentine (1968) as the 'pejorative tradition' and has in Britain found expression in the literature concerned with problems of public health (see, for instance, Wright and Lunn, 1971).

An alternative basic assumption questions the validity of the organic society of shared values and postulates instead an intricate pattern of possibly overlapping sub-cultural systems within a larger society, with more or less sharp boundaries defining their own sets of values. Rex (1961) describes this approach in terms of a conflict model. Sub-cultures often historically precede a dominant culture, as for instance the various regional sub-cultures, or those introduced by immigrant groups; others arise in response to demands of social or cultural structures. Whether a person's life is governed primarily by sub-cultural values, or which aspects of his life are subject to them, depends on the person's sense of identity. Working-class culture has found its most sensitive expression in literature rather than sociological studies, as for instance in Britain by Hoggart (1957) or Seabrook (1967). An extreme interpretation of working-class culture as evident among very poor sectors in western societies has been given in terms of a 'sub-culture of poverty' by Lewis (1967). In contrast to this, Gans (1970) acknowledges typical working-class patterns but denies the existence of a sub-culture of poverty, and explains the regularities of life style observable in certain poverty groups as no more than situational adaptations to constraints imposed by adverse circumstances.

The two basic assumptions - that of the organic society with largely shared values, and that of cultural variety within a host society holding dominant values - also have a bearing on the way in which social mobility is explained. The absence of material wealth, low occupational status, and lack of education, which are characteristic of poverty groups are explained by adherents of the former school of thought as the result of personal failure in a society which offers equality of opportunity, wide choices in occupational career structures, and an effective welfare support system for persons in dependency situations. The alternative school of thought questions the reality of these assumptions and draws attention to an essential aspect of poverty, and that is the absence of political power (Miller and Roby, 1970). Economic deprivation is a source of political deprivation in a society the power structure of which is built on a hierarchy of considerable economic inequality (Goldthorpe, 1974). The arguments supporting this view are based on studies of the incidence, distribution and nature of poverty groups in western industrialized societies.

POVERTY STUDIES

The shift towards the welfare state has meant a recognition of social problems such as poverty and social inequality. Their continued presence in an era of fast rising average standards of living has been instrumental in a re-examination of the validity of the concept of equality of opportunity. The definition of poverty commonly accepted consists of the benefit scale rates variously used in countries which have made basic welfare provisions for the destitute. In Britain the supplementary benefit scale rate is periodically adjusted by Parliament to the cost of living index and to average earnings in industry. The incidence of poverty remains difficult to establish for, as Townsend (1970) has pointed out, income as normally defined in industrialized countries is insufficient to describe economic position: 'In all societies, and particularly those which have large numbers of farmers and smallholders or peasants, there are people with small monetary incomes but substantial incomes in kind and vice versa. There are people with the same wage or salary but widely different income, or income in kind, from employers' fringe benefits. There are people, too, with equal access in principle but not in practice to the public social services' (p. 24). Townsend (1974) suggests that the wider concept of 'resources' should replace 'income' in studies concerned with poverty, and should include current cash income, capital assets, employment benefits including income in kind such as housing or cars, social service benefits, and private income in kind. In a survey of British poverty studies Atkinson (1974) discusses the difficulties of measurement of inequality and of data collection which account for disparities in the estimates that have been made. Nevertheless, an estimate made by the Department of Health and Social Security for the year 1972 will give the reader an indication of the size of the problem of poverty in Britain at the time when our study was made.

TABLE 1.1 Numbers in poverty in 1972

	Households	Persons
Estimated numbers with incomes below supplementary benefit level (excluding recipients of SB)	1,220,000	1,780,000
Estimated numbers with incomes at or not more than 10 per cent above supplementary benefit level (excluding SB recipients)	660,000	1,120,000
Estimated numbers in receipt of SB in November 1972	2,750,000	4,141,000
Total	4,630,000	7,040,000

Source: Correspondence from the DHSS to Brian Sedgemore, MP 13 May 1974, quoted in Field and Townsend, 1975.

A study of relevance to our research is that of disadvantaged children who form part of the National Child Development Study (Wedge and Prosser, 1973). This follows the progress of all children born in one week of April 1958. The number of children in poverty was estimated by identifying children who were receiving free school meals, or whose parents had drawn supplementary benefit during the twelve months preceding the interview. It was recognized that this method of identification would under-estimate the total number of families in poverty, as they often fail to take up benefits or to apply for free school meals through fear of stigma, ignorance or other reasons. Despite this, 14.3 per cent of all children, or in numbers some two millions, were found to be in poverty.

The most important, though numerically not the largest, group in poverty consists of families whose main wage earner is in fulltime work on a wage too low for net income to be equal to entitlement to supplementary benefit. Field (1973) showed that the considerable rise in real earnings, which are estimated to have doubled between 1945 and 1970, has not been shared by workers in low-paid jobs.

The creation in 1971 of the family income supplement to assist families with an earned income below subsistence level recognizes the persistence of extreme inequalities of earnings as well as the inadequacy of family allowances as a means of income redistribution. The second, numerically larger, group in poverty is that of persons in states of dependency; of these sickness, unemployment, disability and old age are the most important. There are two systems of State support for those in dependency situations, national insurance and supplementary benefits. The insurance benefit rates are pegged at a level somewhat beneath the poverty line, although earnings-related additions may bring total benefits above the line for a period not exceeding six months. Low levels of benefit, the contributions conditions, and the short span of time covered in case of unemployment, twelve months in all, as well as gaps in coverage provide the main reasons for poverty. Supplementation on a means-tested basis is not always applied for. There are in all some forty national or local means-tested schemes in operation to relieve poverty, of which family income supplement, supplementary benefit, rent and rate rebates, free school meals, welfare milk, free prescriptions and remission of dental/optical charges are among the most important. Most schemes have take-up rates lower than the number of persons estimated to be qualifying. The family income supplement achieved a take-up rate of about half the expected number of families, and only about 70 per cent of eligible tenants are thought to be in receipt of a rent rebate. The main reasons given for hesitation to apply are dislike of charity, feelings of humiliation and resentment, ignorance, problems with form filling, and bad experiences with earlier claims (Lister, 1974). Another severe weakness in relief measures of this kind consists in the 'poverty trap'. Many low paid workers are made worse off if they increase their income by working overtime or by negotiating a wage increase because of high marginal tax rates, a rise in national insurance contributions, and the loss of means-tested benefits.

The group most vulnerable in terms of sickness or unemployment are manual workers in non-skilled types of employment. A national survey of unemployed men (Daniel, 1974) showed that nearly two-thirds of those out of work were non-skilled even though they constituted only 27 per cent of the workforce. Atkinson (1973) discussed a government study of sickness absence and showed that this was considerably higher for unskilled workers than the average for all workers; the difference was particularly marked among older age groups. At the same time the unskilled are not as well covered by employers' sick-pay schemes as other adult male workers. In conclusion Atkinson refers to the cyclical aspects of poverty (p. 117):

> People may not spend their entire lives in poverty . . . but there are many forces making for the continuity of poverty and its recurrence with predictable frequency. . . . What may appear at first sight to be 'bad luck' is likely to be related to labour market disadvantage.

The recurrence of poverty in a family's life cycle is relevant in relation to savings. The inability of low-income families to accumulate assets of any kind brings about adverse conditions which aggravate their circumstances. Runciman (1974) points out that there are collective differences in typical expectations of life chances, not expressed in terms of income, but in terms of the potential for savings; and he suggests that by this standard the division of households should be 'those for whom the net accumulation of assets is easy; those for whom it's not easy but they manage it; those for whom it's not easy and they don't manage it; and those for whom only a windfall will make it possible' (p. 97). The findings of the First Report by the Royal Commission on the Distribution of Income and Wealth (1975) show a very uneven distribution of marketable wealth among the nation with the top 20 per cent owning 82.4 per cent, and 80 per cent of the nation sharing 17.6 per cent of its wealth (Table 36, p. 89).

HOUSING AND THE DEPRIVED AREA

The disadvantages of persistent poverty result in a weak bargaining position in the housing market. The housing problems of the poor are linked to the shrinking supply of privately rented accommodation; while in the immediate post-war years over 60 per cent of dwellings were privately rented, it was only 17 per cent by 1973 (Eversley, 1975). Low income families are increasingly dependent on local authority housing. The waiting lists remain large, and the number of homeless families continues to grow (Greve, Page and Greve, 1971). It is estimated that in London at any one time a quarter of a million people are under-housed or homeless. However, some overall improvement in the national housing situation has lowered the proportion of overcrowded dwellings from 11 per cent in 1960 to 6 per cent in 1971, and the percentage of unfit dwellings from 12 to 7 per cent (Willmott, 1974).

The process of slum clearance and renewal of areas containing substandard and poor housing has brought with it a sharp rise in mobility of those who are resident in these areas. Eversley (1973), in discussing the housing problems of London, draws attention to this process and the consequent growing polarization in terms of income and ethnic groups. Many of the traditional working-class areas of the city have become reception areas for low-income families, often with many children, who need to be rehoused from slum-clearance areas. This inward movement is balanced by out-migration of skilled and young families, who have the resources to move to the suburbs and the new towns; a phenomenon well known in history but speeded by the arrival in the inner city of families who have little chance to obtain houses on post-war estates. This polarization is of great relevance to those who find themselves in areas of growing deprivation - areas which contain more than a fair share of the unskilled, the unemployed, the disabled, the old and large families, as will be documented later in this book. The figures for unemployment in Tower Hamlets, quoted by Eversley, were three times as high as those in the outer boroughs of London. Similar developments have been recorded in other cities. The areas chosen as locations for Community Development Projects are characterized by populations with low incomes, poor health records, high infant mortality rates, a high dependence on state benefits, and a high proportion of unemployment. Five of the twelve CDP areas reported unemployment rates of between 10 and 22 per cent (Inter-Project Report, 1973).

EDUCATIONAL PRIORITY AREAS

The educational authorities had long been aware of the difficulties experienced by teachers who work in deprived areas. In 1963 the Newsom Council had already drawn attention to the problem in reporting on children of average and less than average ability in secondary schools. 'In certain areas' it is stated, 'schools have more to contend with than the schoolboys' traditional reluctance'; the challenge that teachers have to meet 'comes from the whole neighbourhood in which they work and not, as in most modern schools from a handful of difficult families. The difference is so great as to constitute a difference in kind' (p. 21). The Plowden Council, reporting on children in primary schools in 1967, recommended a concentration of resources in 'educational priority areas'. To assist the identification of such areas the Council suggested that the following criteria should be taken into account: the proportion of unskilled and semi-skilled manual workers, family size, the receipt of state benefits, inadequate housing conditions, poor school attendance, the proportion of retarded, disturbed or handicapped children, incomplete families, and children unable to speak English.
One of the outcomes of the Plowden recommendation has been a policy of selective assistance which included higher capital allocations and equipment grants as well as higher salaries for teachers in schools with 'exceptional difficulties'. These measures were followed by a series of experimental programmes in five areas designated as 'educational priority areas' (EPAs). Halsey (1972), in a report on four of the five EPAs, stated that they suffered from multiple economic and social deprivations which varied from area to area so that diagnosis of the ills of each and prescription for amelioration needed to be based on detailed local studies. In all four areas the children who had been tested for vocabulary and reading at two age levels fell well below the national means. These findings were confirmed by the Bullock Committee (1975) which reported on reading standards and the use of English in schools. While national average standards had been maintained, they masked deterioration of standards among children of unskilled and semi-skilled workers in areas with severe social and educational problems.
The Inner London Education Authority undertook a literacy survey of some 30,000 8 year olds in order to study the relationship between the EPA and reading levels. For this study an index was constructed to identify EPAs (Little and Mabey, 1972); the items forming the index consisted of the Plowden criteria but included two additional ones: pupil turnover and teacher turnover. For purposes of analysis the schools of the ILEA were grouped into four types in terms of degree of educational priority. The relationship of type of school and reading levels of pupils graded by socio-economic group showed a deterioration of attainment with increasing degree of social problems. There was a drop of six points for

children from an unskilled background, ten points for skilled, eleven points for other non-manual, and sixteen points for children from a professional or higher administrative background (Little and Mabey, 1973). In a further study Barnes and Lucas (1975) used a regression analysis to look at the effect of institutional disadvantage on individual pupils and also on groups of children and estimated the effects of family size, occupation of father, school attendance, immigrant status and free school meals in order to gauge the comparative size of the measured institutional effects. Irrespective of schools attended, there was about two years difference in average reading age of disadvantaged and not disadvantaged children aged 8-9 years. The most important finding of this study is that the type of school, expressed in terms of educational priority criteria, has very little effect on the performance of children who come from disadvantaged homes. The more privileged the home circumstances of groups of children, the more their performance is depressed by relative institutional deprivation. The authors conclude that 'policies of positive discrimination which successfully overcame the total effect of school context would do little to reduce the spread of performance among individual children'. They would, in fact, increase the difference between the groups. This is contrary to the expressed hopes of the Plowden Council. If a programme of positive discrimination through schools were successful, the authors argue, then the maximum reduction which would result in the spread of reading scores in the classroom would be around seven points for children of non-immigrant, non-manual workers, but only around five points for children of manual workers, and less than three points for West Indian immigrant children. In other words, the circumstances of the families explain more of the variation in reading performance than do the circumstances of the schools. The implications for educational policy are far reaching.

The concept of the educational priority area has been criticized on other grounds: the heterogeneity of EPAs, both between different priority areas and also within them, the fact that there are more under-privileged children outside EPAs than within them, and that within EPAs there are more children who are not severely socially handicapped than those who are. Barnes and Lucas estimate that on an indicator of cumulative disadvantage there are five disadvantaged children outside EPAs for every two who are in them. Nevertheless, the value of focusing on schools in EPAs lies in an assessment of the institutional problems they generate which are not reducible to the problems of individual children. The school crisis of the inner city finds its starkest expression in manifestations of behavioural malfunction, that is to say, problems of discipline, vandalism and truancy.

BEHAVIOURAL PROBLEMS OF PUPILS

School is the place of the young child's first encounter with representatives from the greater society, the world beyond family and peer-group experience. Little is known about the expectations of teachers in deprived areas concerning behavioural standards of their young pupils. A young teacher, with little or no experience of EPA schools, may have expectations which cannot be met by her socially most handicapped pupils, and she may perceive certain behaviour patterns as defiance or as naughtiness which may be no more than a lack of classroom competence. By competence we mean knowing what being in school is about: not merely learning to read but getting on with the job, not requiring or demanding much supervision, being reliable, being able to organize himself, in short, behaving in an educationally motivated way. The child with severe social handicap is usually quite incompetent in the classroom. What we see as lack of competence may be interpreted by the teacher as 'restlessness', or 'laziness'; the child gives up easily, can never stick at anything, has no persistence, and cannot be relied on. That this is indeed often the case is borne out by an observation made by Midwinter in an account of EPA research in Liverpool: 'Teachers in the area were wont to argue that their colleges had not prepared them adequately for the "cultural shock" of EPA teaching' (Midwinter, 1972, p. 125). We also have some information about social background, place of residence and length of teaching experience in EPAs. Halsey (1972) reports a strong middle-class representation of teachers; two-thirds in Deptford and in Birmingham came from middle-class homes. Very few teachers lived in EPAs. In the West Riding the percentage of middle-class teachers was 41, and only 22 per cent lived close to their schools. In discussing the length of experience among EPA teachers Halsey states (p. 67):

> It seems that the inner ring schools depended to a large extent on young newly qualified teachers without family ties, and that the departure of these teachers either on marriage, the arrival of a first baby, or promotion, lead to a constant turnover of staff.

Do teachers in fact observe and record more behaviour difficulties in EPAs? The research reported by Halsey is not concerned with behaviour problems, but there is research evidence from other studies which suggests a link between behavioural difficulties and increasing social handicap. We discuss this matter fully in chapter 4. While the measures of 'maladjustment' in the classroom differ, the findings of those studies which do record a correlation of 'maladjustment' or 'troublesomeness' and social handicap give information not only about the children's behaviour but also about teacher expectations. They are a pointer towards the main problem with which our study is concerned, and that is what kind of preparation for school do socially handicapped children have? There is little systematic knowledge about what happens to very young children in the milieu of poverty, and how their experiences in infancy and early childhood affect social competence and learning readiness.

The problem is not confined to the classroom; the classroom is no more than the testing ground. The educational system, not only in Britain but in other Western democracies, has been seen as responsible for reinforcing the inequalities that exist in society. There is considerable evidence, some of which is reported in our study, that this process of reinforcement of the handicaps inflicted by poverty is not deliberate. It just happens: children who are not educationally motivated do not profit from what is offered in the schools. It is impossible for the teaching staff to avoid giving these children a feeling that they are somehow lesser citizens, as if they are to blame for their failure. Thus the cyclical process is repeated of their parents' lack of success and adaptation to failure.

The families who worked with us in this study have all had contact with a social services department for advice or assistance during a critical period of their lives; as such they cannot claim to be representative of a large sector of society. But as residents of deprived areas of a city and as parents of large families they share the experiences of more substantial minorities. They have much in common with families of non-skilled workers in general. Poverty is a condition endemic among that sector, especially during that stage of their life-cycles in which they bring up their children.

The debate about educational priorities is already turning towards problems well outside the life of the school. American experimentation with compensatory measures to improve the attainments of disadvantaged children has provided evidence that these children cannot do well at school as long as they are rooted in a milieu of poverty. This, if true, carries a political message. But there is a 'here and now' as well as a goal for the future. Much effort goes into remedial work at school when failure has been diagnosed. Are such remedial measures effective, or are there perhaps other measures more commensurate with the problems of incompetence and lack of educational motivation that should take the place of those currently undertaken in British schools? Our study was not designed to experiment with remedial measures; its main objective is an exploration of the milieu of poverty and the effect this has on child development. Without knowledge of what is transmitted by the family to produce children, who appear to cope well enough in their own environment but who cannot cope in the alien environment of school, it is premature to prescribe alternatives to current remedial practices in schools. The solution to educational failure of disadvantaged children may ultimately lie not in changes in educational methods, but in drastic fiscal measures to bring about a reduction of extreme inequality.

Chapter 2

THE FAMILIES AND THE ENVIRONMENT

Our brief from the Research Unit of the Home Office was for a sociological and psychological study of families known to a local authority children's department (now social services department), to examine the personality development of two children in each family and to relate this to the social and economic circumstances of their homes. It was decided that each family was to have one child aged between 3-4 years who would participate in a nursery play-group, and also a boy of school age who would be the main focus of the psychological examinations.

A descriptive study which proposes to ascertain constant relationships of family circumstances and functional aspects of its members has to be conducted at various levels. In the first place, an account is given of family size and composition, occupations, education, income, housing and the general environment. Our second objective was to assess the school boys' ability, attainments and behavioural characteristics. Each family had a boy either aged 6-7 or 10-11 who was given tests of cognitive development and educational attainments. The older age group was given two self-report scales. Teachers rated behaviour in the classroom of both age groups. The psychological study of these two age groups of children was aimed at making some cautious statements about the effects of four years' schooling and changes in status relative to other children. Although there are national norms for some of the psychological tests which were used, this was not the case for all. We believe it is essential to make comparisons not only with national norms but also with children whose conditions of life are similar in some important respects to those of the main sample, and we chose for controls boys living in the same neighbourhoods and attending the same schools, but less disadvantaged. Criteria for the selection of the two sets of controls for each age group were developed and incorporated into an instrument which measures degree of social handicap. The prevalence of degrees of social handicap in the catchment areas of the children's schools was established by applying the instrument to a large random sample of boys in the same districts. In view of the fact that deprived children frequently suffer from undetected and untreated medical conditions or defects, we arranged that the children were given a general medical examination. The health findings are reported by Dr Mary Brennan (1972).

Each family had a 3-4 year old child who attended a weekly play group. This served a triple function:
1 A service was given to the families during six months of close contact and interviews.
2 Weekly visits to collect and return the children by one of us (HW) established a good relationship; we came to know the families and they came to know us.
3 At play-group sessions the children's social behaviour and their use of play materials was observed, and indices were developed for an instrument by which comparison could be made with another play-group of working-class children who were not socially disadvantaged. The children who attended most regularly were given individual cognitive tests. An attempt was made to sample spontaneous language during play.

Our third major objective was to find out how the families bring up their children. The source of this information is a series of conversations with the families in their own homes covering many aspects of family life. The conversations were guided by interview schedules (Appendix D). Three of a total of six interviews are specifically concerned with child-rearing methods, and the schedules guiding these are adaptations of a set designed by Drs J. and E. Newson. This made limited comparisons with another sample representing all socio-economic classes possible (Appendix E). The cost involved in this part of the work precluded a parallel study among the families of the control boys.

A fourth objective was the ascertainment of delinquency in each family of the main sample, and among the older age group of the controls. Delinquency was studied as a family problem, comparing families who had delinquent children with others who had none and relating the findings to child-rearing methods. The older group of boys, aged 10-11 during the fieldwork, were compared with their controls when aged 12-13 so as to relate delinquency to degree of social handicap.

Fifth, and finally, inter-group comparisons of patterns of child-rearing within the main sample drew our attention to aspects of life in deprived areas which call for specific measures of parenting; but not all parents were sensitive to these aspects. The inter-group differences in parenting have, we believe, a lasting effect on child development.

THE MAIN SAMPLE

The following criteria were adopted for selection of the main sample:
1 residence in old housing in deprived inner city areas;
2 indigenous origin (including Irish);
3 intact family irrespective of marital status;
4 five or more children;
5 one of the children to be aged 3-4 and willing to take part in a play-group;
6 one of the children to be a boy aged either 6-7 or 10-11.

The choice of sex of the school child was made because boys provide a greater proportion of children with educational and behavioural problems, and because the small numbers in the sample would have precluded analysis of sex differences. The two age groups were chosen because the younger boys would be in the upper half of the infant school and at the start of the educational process, and the older group would be in the upper part of the junior school with experience of success or failure over several years. The two ages were sufficiently close to have at least some types of educational tests in common. In addition, several other studies had used these age groups and would provide comparisons with our findings.

The social services department allowed access to records of all families who had approached the department for advice and help, or who had been referred by another agent. Contact with the department was of varying length. The original intention was to select sixty families in all, but sampling on the basis of the chosen criteria presented difficulties, and the final number was fifty-six families, of whom twenty-eight had a boy aged 6-7, and twenty-eight a boy aged 10-11. The procedure was time-consuming, as there were no detailed records of ethnic origin or type of housing, and birth dates of children were recorded only when casework was undertaken. Thus one of us (HW) visited 185 families before the final selection could be made. The sample was drawn in two stages of the fieldwork to enable us to work with smaller groups. Of the 185 families originally contacted, eighty-six had to be rejected because they did not satisfy all criteria for inclusion. Nine families did not wish to take part in the study; another six, who had agreed to co-operate, and whose children had already attended a number of play-group sessions, lost interest after the first or second interview and withdrew. Twenty-two families moved from the areas in which our study was located.

It is the practice of the social services department to refer families in need of long-term supportive assistance, when appropriate, to other social agencies. Two families in the sample had spasmodic contact with the NSPCC (National Society for Prevention of Cruelty to Children), and thirteen were visited regularly by social workers of the Family Service Unit. Another thirteen families were visited by social workers of the department. These families, twenty-eight in all, are described as being on 'preventive supervision'. The other half of the sample consisted of families who had approached the department in a time of crisis for help or advice, and this had been given in an interview or a short series of visits of not more than six in all. As sampling was undertaken by consulting in the first place the records of initial visits, decisions about the type of assistance needed were pending and not known to us. To avoid bias in our attitudes towards the families their positions vis-à-vis social agencies were disregarded until the end of the period of fieldwork. The neat division into two halves, one being supervised, the other having had short contact only, was entirely fortuitous. A comparison between the two categories revealed no significant differences in relation to income, father's work records, family composition, delinquency, or social handicap score (see Table 2.1).

TABLE 2.1 Family characteristics and contact with department

	'No further action'	'Preventive supervision'	Total	p
Family income:				
beneath poverty line	8	12	20	
at or above	20	16	36	ns
Father's work record:				
mainly sick or unemployed	7	12	19	
mainly at work	21	16	37	ns
Family composition:				
8 or more children with 3 or more under 5 years old	5	8	13	
the rest	23	20	43	ns
Delinquent children:				
delinquent families	13	14	27	
non-delinquent families	13	11	24	ns
all children less than 10 years old	2	3	5	
Social handicap score:				
5-7	15	13	28	
8-11	13	15	28	ns

(For definitions of characteristics see text.)

LOCATION

In choosing the older wards of the city as location of residence we had in mind features of educational priority areas which the Plowden Report describes in terms such as this: 'Incessant traffic noise in narrow streets, parked vehicles hemming in the pavement, rubbish dumps on waste land nearby, the absence of green playing spaces on or near the school sites, tiny playgrounds, gaunt looking buildings'. There are in the city many areas similar to this description, but not all have been classified as containing schools of exceptional difficulties, an administrative term used to single out schools in need of special financial support. In making the final choice of areas we were guided by the practical consideration of having an operational radius of not more than five miles from our office and the play centre, and of having clusters of families which would facilitate visiting and collection of children for play sessions. The areas chosen fall within a homogeneous social area. We were guided by the classification of city areas developed at the Centre for Urban and Regional Studies, University of Birmingham

(Edwards, Leigh and Marshall, 1970). Analysis of forty-seven social and economic variables revealed three distinct clusters of which one characterized a deprived area, containing a large proportion of unskilled workers, large families, houses lacking basic amenities, and overcrowding. This cluster is located exclusively in enumeration districts of the more central wards of the city. As the information was based on 1966 census data, we had to avoid some newly developed areas, and also families living in new houses within deprived areas.

Visual impressions of the neighbourhoods confirmed the information obtained by cluster analysis. Almost all the streets in which the sample families lived consisted of houses built in the second half of the last century for occupation by a rapidly growing number of working-class families. The houses were terraced, sometimes grouped around a large communal yard which was reached via narrow gateways cut through a facing row of terraced houses, and many of the houses were built in the back-to-back style. There were few gardens; yards were usually communally used and contained rows of lavatories. Some families fenced off a small patch for play or for growing a few bushes or even vegetables. There were many factories, warehouses, railway sheds and sidings, dilapidated derelict houses waiting for demolition, and open spaces where houses previously stood. In some areas in which redevelopment was in full swing whole streets had been evacuated and houses stood empty, or wide open spaces invited the children to enjoy unlimited boundaries and open skies. The little corner shops in the streets, often run by immigrants, were generally old-fashioned and sometimes very shabby. Public houses at almost every street corner provided the only form of entertainment. Some of the families who took part in our study were within walking distance of a park or a small recreation ground, and some were near a reservoir; others had no such amenities.

HOUSING

All but four of the fifty-six families occupied houses which had been bought by the local authority prior to demolition, then repaired and made habitable for temporary accommodation. Three families rented similar houses from a private landlord. One family was buying the house they lived in, and this was a slightly newer dwelling, built shortly before the First World War. The other houses had been built between 1860 and 1880, and four at the turn of the century.

The absence of basic amenities in a house is a simple way of describing its physical characteristics. If there is no internal lavatory, no fixed bath or shower, no wash handbasin, no hot water supply, the house is classified as sub-standard. The absence of basic amenities usually goes together with a bad state of repair, dampness, absence of natural lighting, ventilation, and poor drainage. There are usually poor facilities for the storage, preparation and cooking of food, and there is inadequate heating. Only five families had a hot water system, and only three had an indoor lavatory.

Overcrowding was measured by means of a Government bedroom standard (Woolf, 1967). This allocates a bedroom on the following criteria:
1 each married couple;
2 each other person aged 21 or over;
3 persons aged 10-20 of the same sex, one bedroom per pair;
4 persons aged 10-20 left over after pairing, paired with a child of the same sex under ten, with a bedroom to the pair, and if no pairing is possible, a separate bedroom for that person;
5 any remaining children under ten paired, irrespective of sex, with a bedroom for each pair.

On the basis of this standard, forty-four families (79 per cent) were found to be overcrowded, seventeen families lacked one bedroom, seventeen lacked two bedrooms, and ten families lacked between three and five bedrooms.

The ratio of numbers of persons per living room indicates the degree of stress the family is subjected to in daily life, especially in cold weather when everybody crowds into the one room with a fire. In many houses the living quarters, in contrast to the bedrooms upstairs, consisted of a single room. The cooker was often in a cupboard or other small separate space with a ventilating duct. But this is not typical of the 'back house' in which cooker and sink are in the living room. The number of persons per living room was two to three for sixteen families (28 per cent), four to five for nineteen families (34 per cent), and six or more for twenty-one families (38 per cent). In view of the fact that one of the purposes of the provision of council housing is to relieve overcrowding, this finding was unexpected. But it points at the plight of large families on low incomes, they cannot afford the rent of better housing commensurate with family size, and they have difficulties in furnishing a larger house and keeping it heated. One family had been rehoused in two old houses which had been connected by an internal door; we found them occupying only one of these houses and keeping only one living room heated. Some of the families had a front room and a back kitchen. The front room was used on special occasions. Some had turned the front room into an additional bedroom.

Another and more disturbing aspect of the housing problem presented by the research sample was the high degree of mobility. This is not the mobility observable in a general population in an era of increasing affluence, it is the mobility of a residual population who have no choices in the housing market. The chronological pattern typical of the families is to start off married life with the in-laws, then a move is made to furnished rooms with the arrival of the first or second baby, and as the number of children grows, further moves are made to find landlords who will take the family. Sometimes there were evictions for non-payment of rents and a stay in a half-way house or in accommodation for the homeless. A substandard house is often offered after homelessness, but sometimes families who find themselves in this type of housing market cannot escape it; when the house is demolished they are offered another short-life house. As the total stock of sub-standard houses is gradually dwindling, more families were given a chance in purpose-built council houses towards the end of our fieldwork. Thirty-three families (59 per cent) had between four and nine addresses in the course of their married lives, thirteen families had three, and only ten had moved not more than once. This pattern of high mobility has adverse effects on social integration and educational development.

FAMILY COMPOSITION

One of the criteria for inclusion in the sample was the presence of both parents. They were not necessarily married nor were the children all necessarily of one father. In fact, three fathers were temporarily absent during fieldwork. Family size ran from five to twelve children, and the older children in some larger families had left home. We divided the families into three groups in relation to the degree of burden experienced by mothers. The families with the greatest parenting burden were those who had eight or more children of whom three or more were under 5 years old. There were thirteen in this group. The families with the least burden were those with no more than five children of whom no more than two were under 5, another thirteen in all (Table 2.2).

TABLE 2.2 Family size and pre-school children

	Total number of children					
	5	6-7	8-9	10 or more	Total	%
Number of pre-school children						
one	6	7	3	–	16	29
two	7	6	1	4	18	32
three	2	1	4	4	11	20
four	3	3	2	3	11	20
Total	18	17	10	11	56	
%	32	30	18	20	100	

Two-thirds of the families have six or more children; one-fifth exceeded ten. We have no records of stillbirths, or children who did not survive. In discussing fertility intentions the following information was given by mothers: 39 considered their families complete, but only 19 of these can be safely assumed to have complete families (12 sterilizations, 4 hysterectomies, 3 menopause). Contraceptive measures were used by 20 (10 intra-uterine device, 10 'the pill'). The remaining 17 mothers (30 per cent) had no reliable safeguards against pregnancy; some expressed emotional dislike of using preventive measures, some expressed a preference for children, some mentioned the observance of Catholic teaching, and some sexual disharmony and an indifference to the problem. With such a large minority antagonistic to family limitation it appears that advice and domiciliary support is needed to overcome specific difficulties. Many mothers expressed a fear of another pregnancy.

PARENTAL EDUCATION AND OCCUPATIONAL HISTORIES

The fathers

Educational experiences varied due to a fairly wide age range from 30 to over 50 years; some of the men had only three years of secondary education, others up to five, but none had stayed on beyond compulsory school age.

Some of the men learnt a trade in the army. At least one man began to serve, but did not complete a period of apprenticeship as fitter-welder as he decided, at the outbreak of war, to join the merchant navy. At the time of interview or prior to unemployment the distribution of occupations, as defined in the Registrar General's 'Classification of Occupations', was as follows: skilled (class III) nine, semi-skilled (class IV) sixteen, and unskilled (class V) thirty-one men. The skilled group consisted of two lorry drivers, previously army mechanics; a fitter-welder; three plasterers and/or painter and decorators; one excavation driver; one factory fork-lift driver; one die caster. The semi-skilled men worked mainly in factories as machine operators, car assemblers, sprayers; some were warehousemen, storekeepers, garage mechanics, and one a concreter. Unskilled men worked mainly on building sites, road works, or as factory labourers. They included one hotel kitchen hand, one road sweeper, one refuse collector, one caretaker, one coal delivery man, and one self-employed rag trader. The men's accounts of previous jobs revealed many changes, sometimes across class. Eleven men from Ireland had worked on the land, and two men from Wales and the West of England had some agricultural experience. Three men could not be interviewed.

The mothers

All had left school at the earliest opportunity. Two Irish women had been allowed to leave school aged 11 and 12 respectively to replace their mothers at home after death. Two women acquired some training in shorthand and typing, and one had some training in nursing. Before marriage the majority of women had worked in factories as capstan, power-press and hand-press workers, or as assemblers or packers. A number of women had been domestic workers in hospitals, hotels, institutions or in private houses. Waitressing and army or navy canteen work were also mentioned. Only two women had been shop assistants. The majority of women gave up work before the first child was born, but some went back to work at various stages. During the period of our fieldwork ten mothers had part-time jobs, working mostly early in the mornings or in the evenings as office cleaners. Two had part-time jobs in factories; one was a hospital orderly and one worked part-time in a dry-cleaners' shop.

Chapter 2

IN AND OUT OF WORK: THE FAMILIES' EXPERIENCE*

The majority of the fifty-six families (thirty-seven) were dependent on the father's earnings for more than half of a two-year period (1969-70) which we surveyed. But the men's actual experience of being in or out of work varied very widely. At one extreme, only three men appear to have been earning throughout the period; at the other end of the continuum, nine families received social security benefits every week for two years.

Our knowledge of these experiences is based upon the information given us (with the permission of the families) by the Department of Health and Social Security (DHSS), the Supplementary Benefits Commission (SBC) and the Department of Employment (DE) on the duration and level of payments of national insurance sickness and unemployment benefit, supplementary benefit and family allowances made to the families. From this information we charted, for each family, the periods during which it received national insurance or supplementary benefits (together called social security). We have to assume that during periods not covered by social security payments the families were dependent upon the men's present or previous earnings (see Appendix A).

To handle this range of experience, we divided the sample into four groups defined by the duration of their dependence on social security benefits:

Group 1: families receiving benefits continuously for 104 weeks.
Group 2: families receiving benefits for seventy weeks or more, but less than 104 weeks. Earnings were thus the principal income source for less than one-third of the period.
Group 3: families receiving benefits for thirty-five to sixty-nine weeks, for whom earnings were the principal income source for between one- and two-thirds of the two years.
Group 4: families receiving benefit for thirty-four weeks or less, or not at all, for whom earnings were the principal income source for two-thirds or more of the period.

The men who were mainly out of work during the period were chiefly those who possessed few employable skills. Table 2.3 shows that the total duration of periods off work in general decreased in relation to the increased possession of skills.

---

* The remainder of this chapter was written by John Veit Wilson, who carried out the analysis of the income sources and levels on which it is based. The terms and methods used, and some of the more detailed findings, are described in Appendix A.

TABLE 2.3 Number of families in each group, by duration of dependence on social security benefits, and social class of fathers

	Social class of father*			Total families	
Group	3 Skilled	4 Semi-skilled	5 Unskilled	Number	% of sample
1 Continuous social security	1	1	7	9	16
2 Mainly off work	-	2	8	10	18
3 Intermediate	1	7	6	14	25
4 Mainly in work	7	6	10	23	41
Total	9	16	31	56	100

* Registrar General's Social Class.

While two out of every five families (twenty-three) had been dependent on earnings for two-thirds or more of the two years, only one-third (nineteen) had been dependent on social security incomes for the same length of time. The remaining quarter of the families (fourteen) had been dependent for about half the time on earnings or social security benefits.

The interruptions of work by unemployment or sickness were long. One man in five (eleven) had average spells off work exceeding three months, and twice as many (twenty-three) had average spells exceeding two months. The duration of spells of dependence on social security benefits is analysed statistically in Table 2.4.

Not only were those mainly unemployed or sick away from work more than twice as frequently as those who were mainly in work, but the mean duration of their spells off work was about four times as long. The intermediate group resembled the group mainly off work in the length of the spells, but the men in this group were off work less frequently. Of the twenty-three men who were mainly in work, only three experienced spells of more than eight weeks on average, whereas all the mainly unemployed or sick and nearly three-quarters of those in the intermediate group had spells of this average duration. The large differences between the mean and median values in certain entries indicate large differences in the experience of different men in that category: some men had a few very long spells off work, whereas others had several much shorter spells. (It should be noted that the row 'Mean duration of spells in weeks' refers to the average time that an individual man was off work; the average lengths of a spell off work, without regard to who experienced it, were 15.3, 10.1 and 4.6 weeks respectively for the three categories. The difference between these and the table entries is another expression of the very different experiences of different men.)

TABLE 2.4 Duration of spells of dependence on social security benefits

Group	2 Mainly off work		3 Inter- mediate		4 Mainly in work	
Number of families	10		14		23	
Total duration of spells in weeks	(70–103)		(35–69)		(0–34)	
Mean	87.4		49.6		12.3	
Median	88.0		47.5		7.0	
Number of spells						
Mean	5.7		4.9		2.7	
Median	6.0		4.0		2.0	
Mean duration of spells in weeks					*	
Mean	19.3		17.8		4.8	
Median	14.6		11.3		4.7	
Men with average spells of more than (cumulative)	N	%	N	%	N	%
8 weeks	10	100	10	72	3	13
13 weeks	6	60	5	36	0	–
26 weeks	1	10	2	14	0	–
39 weeks	1	10	2	14	0	–
52 weeks	0	–	2	14	0	–

* N = 19 because 3 men had made no claims for social security and one man made a single claim that did not indicate a spell off work.

We cannot draw firm conclusions as to whether some of the men were unemployed or sick, since the information on supplementary benefit did not always state if it was paid in supplementation of, or substitution for, national insurance benefits for sickness or unemployment. The records showed that the cash payments of social security benefits had fluctuated for many of the families even during one spell off work. These payments are based *inter alia* on household size and composition as well as, for national insurance, a contribution record. The presence or absence of dependent and non-dependent children, the changing ages of children, the exhaustion of rights to contributory benefits, all gave rise to fluctuations and changing entitlements to national insurance or supplementary benefits calculated on different bases. We can therefore only refer to 'spells off work' without specifying the causes.

Earnings may have been the most important source of income in aggregate, but their receipt or level were no more dependable or constant than different forms of social security benefits. We have shown how frequently and lengthily earnings were interrupted, but they fluctuated even while families depended on them. Men with occasional overtime or bonus earnings or seasonally affected employment; wives with occasional part-time work; people changing jobs - few could rely on an unchanging level of household income for any length of time. This short-term fluctuation of incomes is found among about a quarter of the manually employed population (Sinfield and Twine, 1969). The one constant feature of most of the families' experience was the low level of their incomes. We describe this in the next section.

FAMILY INCOME LEVELS

Compared with the rest of the households in the West Midlands region among whom they lived, the fifty-six families lived on low incomes whether or not the fathers were in work, and these incomes were close to the official poverty line.

To enable us to establish the facts, we complemented our long-term view of the families' dependence on different sources of income by a 'snapshot' view of cash income during a particular and generally comparable short period of time, a week in the middle of 1970 in which the family composition had been as complete as possible (it ranged between March and October but for the majority it was during June or July). We were not able to collect information from which to calculate the value of goods and services received in kind from official or unofficial sources; nor was the information on the value of rebates or charges remitted sufficiently reliable or standard for it to be included. The study therefore refers only to the principal sources of cash incomes - earnings or social security, together with family allowances. (Further details of the sources and limitations of the data are given in Appendix A.)

While the West Midlands average household income in 1969-70 was £34.08 per week, Table 2.5 shows that thirty-three of the thirty-seven families dependent on earnings for more than a third of that period (groups 3 and 4) had average household incomes when the fathers were in work of £25.63. These incomes consisted of take-home pay and family allowances, plus - in only three cases - wives' earnings. The thirty-three families (in groups 1, 2 and 3) dependent on social security sources of income for more than a third of the period had household incomes which averaged £18.93 when the fathers were off work. Thus when the sample families were out of work their incomes were only about half of those of households in the region. Even when the families were in work, their average incomes rose to only three-quarters of the regional average.

The household incomes of the men in work were low because their earnings were low. The distribution of the men's gross earnings is shown in Table 2.6.

The median gross earnings of the thirty-three men for whom we have information were only three-quarters of the median for the same class of manual workers in the region in which they lived and with whom they should be compared (since wage rates are affected by local labour markets, national averages may conceal wide regional or local variations).

Chapter 2

TABLE 2.5 Household size and household income

Group	1	2	3	4	Total
Numbers of families	9	10	14	23	56
Mean h/h size:					
Sample: Parents	1.2*	2	2	2	1.87
Sample: Dependent children mean	6.1	6.3	6.3	6.3	6.27
Sample: Dependent children median	5	6	6	6	6
Sample: Total mean size	6.3	8.3	8.3	8.3	8.14
West Midlands Region					2.98
Mean h/h income: £/week				**	
Sample: Groups off work	17.66	19.27	19.51	–	18.93
Sample: Groups in work	–	–	25.66	25.61	25.63
West Midlands Region					34.08
Per capita income: £/week					
Sample: Groups off work	2.42	2.32	2.35	–	2.36
Sample: Groups in work	–	–	3.09	3.04	3.06
West Midlands Region					11.45

Sample as % of West Midlands Region:	Families off work N = 33		Families in work N = 33+	
Household size	263		279	
Household income	56		75	
Per capita income	21		27	

* 7 families were paid social security as for single-headed households.
** Information on earnings available for only 19 men of the 23 in group 4.
Source of West Midlands data: Department of Employment. 'Family Expenditure Survey 1970'. HMSO 1971, Table 26 – Average total weekly household income; Table 43 – Average number of persons per household.

TABLE 2.6 Distribution of gross weekly earnings of men in groups 3 and 4

	Men aged 21 and over, full time, paid for a full week					
	1969-70		April 1970			
	33 men in sample		West Midlands		Great Britain	
	£	% of median	£	% of median	£	% of median
Highest decile	27.40	131	39.90	143	37.70	147
Upper quartile	23.00	110	34.00	122	31.30	122
Median	20.90	100	27.90	100	25.60	100
Lowest quartile	17.00	81	22.80	82	20.80	81
Lowest decile	15.00	72	19.00	68	17.20	67
Average	20.95		28.90		26.80	
Sample median as per cent of medians of:		100		72		78

Source of Regional and National Data: 'Department of Employment Gazette', Vol. LXXIX, No. 1, January 1971. New Earnings Survey 1970 Part 3 - Analyses by Region. Tables 54 and 58, giving details of gross weekly earnings of full time male manual workers aged twenty-one and over, paid for a full week, for April 1970, calculated on Basis D.

The men in the sample appeared to have a shorter as well as a lower range of earnings than the regional and national averages, with the top decile less than one-third above the median, compared with over two-fifths regionally and nearly one-half nationally. The highest reported gross earnings of the men in the sample were £34.00 a week, and the lowest £14.10.

There was not much difference between the earnings of the men who spent more than two-thirds of the survey period in work (group 4) and those who spent only about half of it in work (group 3). The mean earnings of the nineteen men in group 4 was £21.04, and of the fourteen men in group 3 was £20.82 a week, the average of both groups being £20.95. This was well below the regional average of £28.90 or national average of £26.80 at that time.

'Low pay is a relative concept', said the National Board for Prices and Incomes in 1971. 'We have identified as areas of low pay those where earnings fall into the bottom tenth of the "earnings league"...' (National Board for Prices and Incomes 1971). Whatever the limitations of this definition, it is notable that no less than one in every four of the thirty-three men in the sample had earnings the same as or less than the lowest tenth of adult male manual workers nationally. By regional standards these fathers were even worse off, since four in every ten (thirteen of the thirty-three) had earnings of £19 or less, compared with one in ten regionally.

But the situation was worse than these regional comparisons make it appear. The average income per head of each member of the sample families (as shown in Table 2.5) was only between one-fifth and one-quarter of the per capita household income of the West Midlands population. This was because the average size of the sample households (including their average of over six dependent children) was more than two-and-a-half times larger than the regional average of just under three persons.

These low levels of per capita income brought the families' incomes down to the poverty line. The official poverty line is based upon the supplementary benefit scale rates, which allow certain sums of money for each dependent member of a household according to his or her status, and age if below eighteen. Allowance is also made for specified housing costs. We do not believe that the poverty line can be defined satisfactorily in terms of the SBC's scale of financial 'needs', since sociological research shows that poverty, if defined as the income level correlating with the inability to attain culturally prescribed levels of social experience or performance, occurs at much higher income levels (Townsend, 1970). The scales are, however, the prevailing official measure of 'minimum income needs' sanctioned by Parliament. They thus provide a yardstick against which to measure if people would officially be regarded as in poverty. In our calculations we have therefore used the same measures of poverty, involving slight modifications of the SB scales, as were used by the DHSS in their study of Two Parent Families (Howe, 1971: further details in Appendix A).

As the intermediate group of fourteen families had experienced comparable periods of dependence on earnings or social security, we compared the levels from each source in mid-1970. The income per capita of the members of these families was on average a third higher when the father was in work than when he was not (see Table 2.5). Comparing their household incomes (see Table 2.7), two families were marginally worse off in work, the remainder somewhat better off.

All but two of the fourteen families had incomes below the official poverty line when they were dependent on social security. Four had incomes below the poverty line when the fathers were in work and a fifth would have fallen below the line had the mother not worked part-time as well. Two of these four families, one containing eight children and the other six, had incomes even further below the poverty line if and when the father could find work; one of them was wage stopped, but not sufficiently to reduce his out of work benefit below his net earnings.

The remaining twelve families in this group were better off when the father was in full-time work than when he was out of work. Including the two families with incomes below the poverty line in work, their average household incomes were about 40 per cent above the levels of their incomes when they were dependent on social security sources. This may have represented a considerable benefit to them, but it did not remove them from the levels of living of the poor in Britain. Even when the fathers were in work, their household incomes were on average not much more than a quarter above their poverty lines. This experience of low levels of earned household

TABLE 2.7 The relation between household incomes from social security sources and from earnings of families in group 3 (Intermediate)

Part A	Income level when it was mainly from:	
	Social security sources	Earnings
Number of families	14	14
'Poverty line' £/range	15.64-26.70	16.14-27.20
Mean	19.95	20.45
Median	19.35	19.85
Income £/range	15.40-26.25	18.00-35.90
Mean	19.51	25.66
Median	19.07	23.40
Income as per cent of poverty line		
Range	90-112	85-217
Mean	98	128
Median	98	129

Part B	Families with incomes higher from:	
	Social security sources	Earnings
Number of families	2	12
Cash difference £/week	(less than earnings)	(more than soc. sec.)
Range	1.63-2.35	1.65-19.54
Mean	1.99	7.50
Median	–	6.45
Income in work as per cent of income from social security		
Range	88-93	106-219
Mean	91	141
Median	–	140

TABLE 2.8 Income received from social security sources by groups 1-3 and its distribution above and below the poverty line

Group	1 Continuous social security	2 Mainly off work	3 Intermediate
Number of families	9	10	14
Poverty line:			
£ range	12.60-29.70	16.36-24.03	15.64-26.70
Mean	17.61	19.98	19.95
Median	15.35	19.35	19.35
Total income:			
£ range	13.38-29.00	16.67-24.28	15.40-26.25
Mean	17.66	19.27	19.51
Median	15.65	18.62	19.07
Income as per cent of poverty line:			
range	91-115	85-102	90-112
Mean	100	97	98
Median	99	99	98
Families with incomes *on or above* poverty line by £:			
2.00-2.99	1	-	1
1.00-1.99	-	-	-
0.00-0.99	2	5	1
Total *on or above* line	3	5	2
Families with incomes *below* poverty line by £:			
0.01-0.99	4	2	10
1.00-1.99	2	2	1
2.00-2.99	-	-	1
3.00-3.99	-	1	-
Total *below* line	6	5	12

incomes was shared by the families in work for two-thirds or more of the two years. Only one of the nineteen families in group 4 whose earnings we knew had an income below the poverty line when the father was in work. But this did not mean that the range of earned household incomes was higher than where the fathers had longer or more frequent spells off work: the mean income as a percentage of the poverty line was 128 in group 3 and 124 in group 4; the medians were respectively 129 and 123, within a range of 85-217 in group 3 and 88-173 in group 4.

The income from earnings of only seven of these thirty-three families (21 per cent) fell at or above the 140 per cent of the SBC scale rates which Abel-Smith and Townsend took in 1965 as their definition of poverty because 'we found that there were many households receiving national assistance whose level of living was 40 per cent or more above the basic rates' (Abel-Smith and Townsend, 1965).

Moreover, had the families in our sample not been paying exceptionally low rents (at £1.42 a week on average they were exactly half of the regional average for council houses - see reference on Table 2.5 for regional data source) their poverty lines would have been higher and their incomes in work would have been even closer to poverty.

When they were dependent on social security incomes more families fell below the poverty line than rose above it. Three out of every five (thirty-three) of the fifty-six families had been dependent on social security for eight months or more in the two years. Table 2.8 shows the relation between the poverty line and the social security incomes these families were receiving in mid-1970. 'Only two families had cash incomes more than £1.00 above their poverty lines, and two had incomes falling more than £2.00 below the line.

The duration of dependence on social security benefits made little difference to the incomes the families received. If one allows for the fact that seven of the nine families receiving continuous social security incomes were treated by the SBC as single headed, then not only the average poverty lines but also the mean and median incomes of the three groups were similar, and the latter were at or below the poverty lines. This is inconsistent with the common assumption that the payment of the long term addition (see Appendix) of 50p. a week (the level in 1970) to those on supplementary benefit but not in the labour market, and of earnings-related supplements in addition to national insurance, would lead to incomes above the basic rates. But the men in the sample had rarely had annual average earnings high enough to entitle them to the latter when they were sick or unemployed (none were reported in mid-1970). The details of the additions to and deductions from the social security scales and the net effect they had on the families' incomes are given in Appendix A. It shows that the majority of families had more deducted than added to their scale incomes from supplementary benefit.

Hence the most important reason why most of these thirty-three families lived below the poverty line while maintained by the State was because it appeared to pay them less money than was paid on average to families receiving supplementary benefit nationally, or simply less than they seemed to be entitled to get. In investigations of the reasons for poverty this causal factor is rarely mentioned: we believe it deserves more attention.

Chapter 3

THE CONTROL GROUPS AND THE RANDOM SAMPLE

The investigation of cognitive, behavioural and social development of the boys of the main sample was to be compared with national norms whenever these were available. But such an investigation calls for more meaningful comparisons. We decided that the more appropriate reference population consisted of children from the same schools, of the same ethnic origin, living in the same neighbourhoods and sharing the same material environment. The choice appeared to be between random sampling and sampling by criteria which would provide contrasts. The former would have the advantage of discovering the distribution of characteristics among children in the area against which the boys from the main sample could be described. The latter would make comparison possible between two sharply contrasting groups of children of, for instance, 'functioning' and 'malfunctioning' families. The final choice was the selection of two stratified comparison groups to combine the advantages of the two alternatives.

The principle of selection of the control groups rests on the concept of 'social handicap'. The focus boys are drawn from families who can be described as severely socially handicapped, even though the families were not selected with reference to this concept. It is not our intention to define the concept 'social handicap' in this context except in operational terms, but we feel it is important that we should declare our point of view. The expression 'social handicap' implies a causal link between a child's underfunctioning and some constant feature of his environment. This implication is intended. Certain environmental features are known to be related to under- or malfunction. Even if a strongly hereditarian viewpoint were taken, a considerable proportion of the variance is admitted to be due to environmental factors. Thus the concept of social handicap emphasizes environmental factors rather than genetic or constitutional factors; but this does not imply that the difference found in groups representing different degrees of social handicap are entirely due to social variables.

A measure of social handicap is not a direct measure of the features of the cultural environment that are thought to have critical effects on intellectual and personality development. It merely aims at establishing degrees of disadvantage. Its effectiveness is to be judged by its success in doing so. The choice of items for the instrument was made on the basis of sociological data used in other studies concerned with educational performance. Most of these make use of parental occupation and parental education, and these two factors discriminate if the sample is drawn from a general population, but they fail to do so among a relatively homogeneous population in a deprived area. Family size is known to have a bearing on performance: children from large families, on average, score less well on ability and attainment tests than children from smaller families. Furthermore, as shown by Douglas (1968), the effect of family size is aggravated by inadequate home conditions and poor schooling. West (1969) found social handicap to be correlated with misconduct of boys in schools; social handicap was scored on seven items - inadequate family income, social class V, very unsatisfactory housing, six or more children, children physically neglected, interior of house very neglected, and family supported by a social agency. Askham (1969) suggests that class V, unskilled manual workers, may be grouped into 'stable' and 'unstable' families on such criteria as long-term or frequent unemployment and absence of male head of household. The two groups were found to differ significantly on areas of residence, on numbers of children in the family, and on children's intelligence test results. Birch, Richardson, Baird, Horobin and Illsley (1970) found that subnormal children from classes IV and V with an IQ of sixty or greater and with no clinical evidence of central nervous system damage were over-represented in families with the following characteristics: five or more children, residence in inter-war tenements, crowded housing conditions, and mothers who, prior to marriage, had been engaged in non-skilled manual occupations. They also found an association between these risk conditions and low intellectual functioning among siblings of minimally subnormal children in classes IV and V. Although published at a date after the start of our study and thus not directly related to our thinking, the National Child Development Study's publication by Wedge and Prosser (1973) confirms the close link between poverty, poor housing, large family size and educational attainments.

Thus there appears to be a cluster of sociological variables which, in addition to occupational and parental educational classifications present degrees of social disadvantage. They are size of family, per capita income, area and condition of housing, regularity of father's employment, and physical condition of the child. The selection of items for a social handicap instrument was determined by the intention to design the instrument for use by teachers. Per capita income appeared to be an important variable; West, in fact, considered poverty to be the main single predictor of behavioural problems in children. But teachers have no knowledge of parental incomes. Three pieces of information, closely related to poverty, were ascertainable: father's social class, size of family, and adequacy of child's clothing. These formed the first three items. In addition, it was decided to include an item measuring truancy or unexplained absence from school. Earlier studies by Wilson (1962), Philp (1963) and Tyerman (1968) had shown the high incidence of absence from school for reasons other than health among the poorest families. The Department of Education and Science does not provide information on this subject, but many local authorities would admit informally that truancy rates are related to areas, and deprived areas tend to have an undue share of truants. Finally, in response to the Plowden findings on parental interest in education and children's attainments, we included an item describing degree of parental contact with school. It had the lowest weighting of one point as it was not found to be very discriminating. In a random sample of 389 boys from the catchment area of the research only 10 per cent of parents were rated by teachers as 'taking the initiative in discussing the child's progress' or as 'supporting the school's efforts in an active way'. The percentage is somewhat higher for infant school children than for the juniors. Related to social class, contact with school does not vary markedly: only about 15 per cent of skilled workers' wives were rated as being actively involved. We wish to disclaim that this information is a valid measure of parents' interest in their children's education. We found in parental interviews that difficulties in communication and social insecurity play a large part in determining contact between home and school.

The social handicap instrument is thus made up of weighted scores from the following five variables: social class, family size, school attendance, adequacy of child's clothing, and parental contact with school. A variable relating to quality of housing was not included since the study was located in deprived areas. If the catchment area had been a different one, such an item would have had to be included. The addresses of all control boys were checked later and found to be in enumeration districts which, on the basis of the cluster analysis described earlier, scored above the mean of cluster I, in other words, they had large concentrations of factors defining a poor, generally decaying and deprived area.

Maximum weight of three points is given to poor attendance at school (unexplained), and inadequate clothing, two factors that indicate most clearly that the child is socially handicapped. Two points were given to class V and to family with five or more children. One point was given to parents who were only seen at school in moments of crisis if at all. Each boy's class teacher was given written instructions defining clothing, school attendance and parental contact. Class and family size were ascertained by the research team from the school's record cards, and percentage attendance over two terms was entered by the research team from the school's attendance register. Total scores ranged from one to eleven. Among the higher scores will be more children with poor clothing, much absence for suspect reasons, from large families, and unskilled fathers. The weighting system is designed in such a way that scores of five or more points will always include boys scoring heavily on inadequate clothing and/or attendance. Thus in a simple way the scale identifies children who are showing severe effects of material and social deprivation. On the other hand, the children who scored adversely only on size of family and social class are most likely in the range of scores three to four. This range we have called 'moderate social handicap'. One or two points, or none at all, classify the child as having a 'low social handicap'.

SELECTION OF CONTROL GROUPS

A pilot study indicated that the 'low', 'moderate', and 'severe' social handicap groups would account for about 40 per cent, 30 per cent, and 30 per cent respectively of the indigenous boys in the catchment areas of the research. It indicated also that the boys from the families known to the social services department were likely to score five or more. On the basis of these findings it was decided to select two control boys for each focus boy in each age group, one scoring 0-2, and the other 3-4. A list of twelve boys, including the focus boy, was compiled for each boy with an age range of six months on either side of the focus boy. The class teacher was asked to rate the twelve boys for reasons of absence, adequacy of clothes and parental contact with school. The identity of the focus boy was not disclosed. Social handicap scores were assigned to all after ascertainment of attendance and parental occupation, and final selection was made from the boys nearest in age, one from each of the two social-handicap groups. Immigrants were not included. Each control group should have been exactly equal in numbers with the focus boys in each age group; in fact, due to re-scoring father's social class, two boys (already tested) had to be upgraded, making the final numbers 26 and 30 for the 28 focus boys aged 6; and in the age group 10, 5 'spares' swelled numbers to 30 and 31. Thus there were 117 controls in all for 57 focus boys. (One of the research families had a 6 and a 10 year old boy.)

THE RANDOM SAMPLE

At a later stage of the research it was decided to test the indications of the pilot study by obtaining social handicap scores from a larger random sample of 389 boys attending the schools used in the study, at ages 6 and 10. Ten typical schools were selected, and the sampling aimed to be exhaustive in these: all boys, non-immigrant, of the two age groups were included. The methods used for obtaining the social handicap scores were the same as those used previously, but if any item of information could not be ascertained, the boy was dropped from the sample. The frequency distribution of items was used to check the representativeness of the control groups, and to indicate how far the focus groups were different from a randomly selected severe social-handicap sample. The results showed that in fact the distribution of social handicap in the catchment area was 40, 30 and 30 per cent respectively; that the control groups were in general representative of their strata in these areas; and that the focus groups, as expected, tended to be skewed towards the severe half of their handicap group, although the tendency is not statistically significant. Thus our assumption that families known to a social services department would be severely socially handicapped proved to be correct.

The representativeness of the samples as families typical of deprived areas - not including immigrants - is confirmed by the distribution of social class and family size for each of the two age groups. Table 3.1 shows that in fact class and family size are distributed similarly in the random sample.

TABLE 3.1 Random sample of 389 boys by age group

Social class	Frequency distribution by social class (per cent)					
	II+III N-M	III M	IV	V	Unemployed	Not known
6-7 years	4.3	44.8	28.8	14.1	6.7	1.2
10-11 years	4.9	39.8	31.4	17.3	4.4	2.2

	Frequency distribution by family size (per cent)										
Family size	1	2	3	4	5	6	7	8	9	10	11+
6-7 years	3.7	17.2	13.5	20.9	14.1	11.0	6.1	6.1	2.5	3.7	1.2
10-11 years	3.1	15.9	16.4	16.4	15.9	12.8	5.8	6.6	1.3	3.1	2.7

As regards family size, an initial assumption would be that the 10-year olds came from larger families. However, the distribution is very similar at the two ages and the median of four is the same. If one assumed that the older boys should tend to come from larger families, the result could be thought to show poor sampling. In fact we believe that the 10 year old boys in the areas under study do not come from larger families. Our own focus sample tended to have children closely spaced during a period of uncontrolled fertility, and when reaching a family size considered to be a stress situation the families practise more effective control. Sterilization, the intra-uterine device and the pill play a part at this stage. In fact families with four or more children are almost evenly distributed between the two age groups with 66 and 64 per cent respectively. Another not incompatible explanation of the equally large families in both age groups is the effect of per capita income on the ability to choose housing: the smaller families which one might expect among the younger boys are not heavily represented because such families are more often able to afford better housing outside the inner city area while it is undergoing redevelopment. The preponderance of the large family in this sample is striking. Douglas (1964) found that children in families with five or more children form less than a quarter of a national 'lower working-class' sample consisting of manual workers and their wives, neither of whom had further education or came from a middle-class home. In contrast, 47 per cent of the random-sample boys came from families with five or more children, a difference which is significant at well beyond the 0.001 level.

The overall picture of the distribution of social handicap among a school population in the central, deprived areas of a large city can be seen in Table 3.2. The distribution of social class in the whole sample is as expected; practically half of the fathers are

TABLE 3.2 Random sample: percentage distribution of categories in social handicap items

		Total score category and age group								
		0-2 = Low		3-4 = Moderate		5-11 = Severe		Whole sample	
	Age	6-7	10-11	6-7	10-11	6-7	10-11	6-7	10-11
Father's	skilled and upwards	70.4	69.1	27.0	28.8	37.7	27.9	49.7	45.7
occupation	semi-skilled	28.2	29.8	35.1	45.5	26.4	21.3	29.2	32.1
	unskilled *	(1.4)	(1.1)	37.8	25.8	35.9	50.8	21.1	22.2
No. of children	1-2	43.6	42.6	5.3	4.5	1.9	0.0	20.9	19.0
in family	3-4	35.2	38.3	47.4	29.9	26.0	27.7	34.4	32.7
	5-6	12.7	11.7	28.9	44.7	38.9	36.9	25.2	28.8
	7 +	8.5	7.4	18.4	20.9	35.2	35.4	19.6	19.5
Clothing	Good	53.5	71.3	21.1	29.8	3.7	4.6	30.1	39.8
	Adequate	46.5	28.7	78.9	68.6	18.5	38.5	44.2	43.4
	Poor	0.0	0.0	0.0	1.5	77.8	56.9	25.8	16.8
School	> 90 per cent	78.9	70.2	55.3	62.7	27.8	35.4	56.4	57.5
attendance	75-90 'genuine'	16.9	28.7	21.0	23.9	25.9	9.2	20.9	21.7
	75 - 'genuine'	1.4	1.1	5.3	4.5	5.5	4.6	3.7	3.1
	75-90 'suspect'	2.8	0.0	13.1	8.9	9.3	18.5	7.4	8.4
	90 - 'suspect'	0.0	0.0	5.3	3.3	31.5	32.5	11.7	9.3
Parental contact	'Much'	23.9	14.9	10.5	6.0	1.8	0.0	13.5	8.0
with school	'Some'	64.8	62.8	28.9	34.3	31.5	12.3	45.4	39.8
	'Little'	11.3	22.3	60.5	59.7	66.7	87.7	41.1	5.2
N =		71	94	38	67	54	65	163	226

* Includes chronic unemployed; the 'unknowns' have been distributed proportionately in this table.

in semi-skilled or unskilled occupations, and the percentage of 'white collar' occupations is negligible. In the 'low social-handicap' group only about 1 per cent of children come from families where father is in non-skilled work; these families must be small. Average size of family increases, of course, with increasing degree of social handicap (by definition) as can be seen in Table 3.3. But it must not be overlooked that about 20 per cent of the 'low' group and about 50 per cent of the 'moderate' group have families of five or more children.

TABLE 3.3 Random sample: mean family size and social handicap

	Mean family size	
	age 6-7	Age 10-11
Social handicap		
low	3.5	3.4
moderate	4.7	5.3
severe	6.1	6.0
All families	4.6	4.7

Poor clothing is almost entirely associated with one or more other handicapping features, as only 1.5 per cent appears in the 'moderate' group, the rest in the 'severe' group. Another aspect which deserves comment is the generally low level of parental contact with schools. Only about 10 per cent of parents were rated as having contact, and the percentage is somewhat higher for the 6-7 year olds than the 10-11 year olds. This trend is maintained through the three social handicap groups, with the parents of the younger children taking a more active interest in school affairs. But even at the level of 'low' social handicap, parents, the majority of whom were in class III, did not have much contact. Apart from our own data we have been unable to find material for comparison, but we believe that this measure is a distinguishing feature of the population under study. The matter was discussed with mothers of the main sample and their views are reported in a later chapter.

The main interest in testing the goodness of the social-handicap instrument lay in the answer to the question whether there are trends associated with degrees of social handicap when testing ability and attainment at school or when rating behaviour in the classroom or elsewhere. It must be remembered that the population is almost entirely working class, family size, being on average somewhat over four, is well above national average, and the environment in which the families live is the poorest of a large city. Even within a relatively homogeneous population such as this, is there evidence of a consistent relationship between degree of social disadvantage and children's cognitive and behavioural development? These questions guided us in our work.

Chapter 4

THE BOYS AT SCHOOL

This chapter presents the results of psychological tests and ratings on the 6-7 and 10-11 year old boys from our main research families and on their control boys of low and moderate social handicap. Some readers may prefer to read these results in summary form in chapter 12, since the present chapter is somewhat more technical than other parts of this book. Also, chapter 12 gives the discussion of the data, so our present purpose will be mainly to give the results themselves.

COGNITIVE TESTS

We will first present the results of the major cognitive tests, those with norms on national and, if possible, local samples. The full details of these and other tests are in Appendix B. At each age the major tests consisted of verbal (vocabulary) scales, non-verbal tests, and measures of reading. All tests were individually administered in school, and none of the verbal or non-verbal scales required the child to read. One factor influencing the selection of some tests was the availability of data on broadly representative samples in the West Midlands as well as national norms.
The verbal measures were:
6-7 years   Mill Hill Vocabulary Scale (Raven, 1958).
              English Picture Vocabulary Test 1 (Brimer and Dunn, 1963).
10-11 years WISC Vocabulary Subtest (Wechsler, 1949).
              English Picture Vocabulary Test 2 (Brimer and Dunn, 1963).
Vocabulary Scales are usually highly correlated with measures of general verbal ability.
The non-verbal measures were:
6-7 and 10-11 years: Ravens Coloured Progressive Matrices (Raven, 1958).
10-11 years WISC Block Design Subtest Wechsler, 1949).
            WISC Coding Subtest.
Some of these measures (Raven's, Block Design) were chosen to measure non-verbal reasoning, and the others are more heavily loaded with

'lower level' perceptual-motor skills. It is commonly thought that the disadvantaged do better in the non-verbal than the verbal area. The reading tests were:

6-7 years   Southgate Reading Test 1 Form C (Southgate, 1959).
10-11 years   NFER Reading Test AD (Watts, 1958).
              Vernon Graded Word Reading Test (Vernon, 1938).

All these reading measures are well known.
The histograms in Figures 1-5 and Figures 6-13 summarize the pattern of mean scores on these scales for the 6-7 year olds and 10-11 year olds respectively, converted to a national score base with a mean of 100 and standard deviation of fifteen points. Where the original scores were not in such a base (Ravens and Southgate at 6-7 years; WISC subtests, Ravens, and Vernon at 10-11), the conversions must be regarded as approximate but good enough for our purposes. The original data are in Appendix C, Table C.1. In the identification of each bar on the histograms, 'Nat.' is the mean score for the best national standardization sample which was available; 'Mid.' refers to the mean of local Midlands data from representative samples; 'B + G' indicates a sample of both sexes; 'B' stands for one composed of boys only. For details of the national and local samples used, the reader is referred to Appendix B, where it will be seen that for some scales the national mean was derived not from the original published norms but from other data considered more appropriate because it was British rather than American, or referred to boys only.

Using these results, we may attempt to examine the standing of our sample in relation to national and (where available) local norms. The expectation might be that the inner-city school population is generally below average on tests of ability and achievement. The concept of an Educational Priority Area as put forward in the Plowden Report (1967) might reinforce this expectation by giving the impression that such zones of cities are rather homogeneously below average, although the Report itself does not go into the question. There is work from the Schools Council Project on Compensatory Education (Ferguson, Davies, Evans and Williams, 1971) indicating considerable inter-school variation inside 'deprived' areas. If we link the possibility of such variation with social differences in the schools' catchment areas, and also if we consider that the Low Social Handicap Controls (LSH) are heavily composed of social class III, the largest single proportion in any standardization sample, then we should find that the LSH group are near the national or local average for their age. The Moderate Social Handicap Controls (MSH) and the focus groups should be progressively below average.

It will be seen from the histograms that the LSH groups at both ages do indeed have means quite near (not significantly different from) those of both national and Midlands samples on all tests except reading. The poor showing of even this, the least disadvantaged group, on reading may seem surprising in view of their scores on the tests of verbal and non-verbal ability. In these ability measures their mean corresponds to IQ scores in the high 90s on the usual basis of a national mean of 100 and standard deviation of fifteen points. Thus their means on these tests are firmly around the middle of the scores from representative samples, yet their

Chapter 4

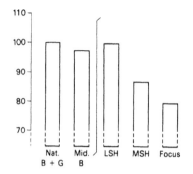

Figure 1 English Picture Vocabulary Test 1: sample means, 6-7 year olds

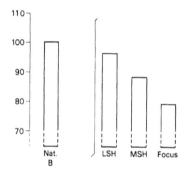

Figure 2 Mill Hill Vocabulary Test: sample means, 6-7 year olds

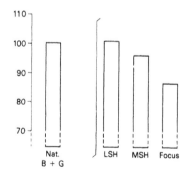

Figure 3 Raven's Coloured Progressive Matrices: sample means, 6-7 year olds

43   Chapter 4

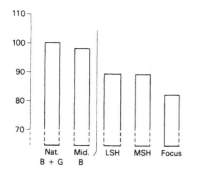

Figure 4   Southgate Reading Test 1, Form C: sample means, 6-7 year olds

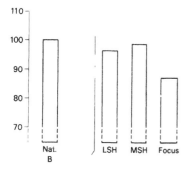

Figure 5   Draw-a-Man Test: sample means, 6-7 year olds

Degrees of freedom: Overall 2, 81
                       Comparisons between pairs of means 1, 81

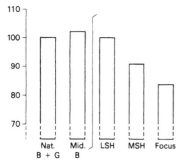

Figure 6   WISC Vocabulary Subtest: sample means, 10-11 year olds

44 Chapter 4

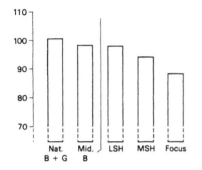

Figure 7 WISC Coding Subtest: sample means, 10-11 year olds

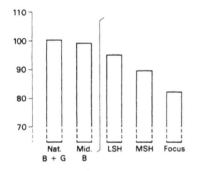

Figure 8 English Picture Vocabulary Test 2: sample means, 10-11 year olds

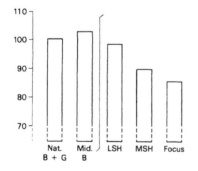

Figure 9 WISC Block Design: sample means, 10-11 year olds

45   Chapter 4

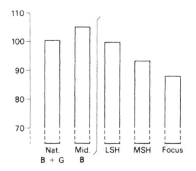

Figure 10   Raven's Coloured Progressive Matrices: sample means, 10-11 year olds

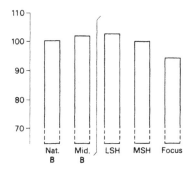

Figure 11   Draw-a-Man Test: sample means, 10-11 year olds

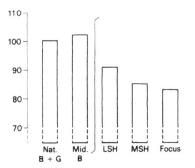

Figure 12   Vernon Graded Word Recognition Test: sample means, 10-11 year olds

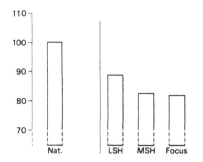

Figure 13 NFER Sentence Reading Test AD: sample means, 10-11 year olds

Degrees of freedom: Overall 2, 87
                     Comparisons between pairs of means 1, 87

reading scores seem to be in the lowest quarter. We must, therefore, confront the possibility that this indicates some specific deficit in reading in the inner-city schools we sampled. This requires careful examination, because it may be interpreted as indicating some deficiency in the teaching of reading in the schools contributing subjects to the study. We can only begin to consider such an explanation if the control samples are not in some way biased towards the selection of poor readers. At the younger age, no such bias seems possible, apart from some undetectable sampling error due to small numbers. At the older age, the control groups show what may be an excess of B stream children in those schools which were streamed (half the sample were in unstreamed schools, and the remainder were in schools with two streams according to ability). The instructions for choosing the control subjects specified that care should be taken to select from the entire range of boys in a school, since taking controls from the same class as focus boys could well result in an excess of B stream controls (on the reasonable expectation that the focus boys would often be in the lower stream).

Presumably due to some defect in random sampling, the controls (LSH and MSH) were almost equally distributed over A and B streams, when there should presumably be an excess of A if one considers that these boys come from the least socially disadvantaged indigenous families in areas with about one-third West Indian or Asian children. We have performed calculations to estimate the average effect of the excess of B stream boys in the combined control groups at 10-11 years (Phillips, Wilson and Herbert, 1972, Appendix D), and it seems to be only in the region of one or two points on the basis of standardized scores which have a standard deviation of fifteen. If our estimate is right, then even if the stream-sampling bias were corrected, the LSH and MSH groups would remain well behind national and indeed local reading norms at 10-11 as well as at 6-7 years. This possible reading deficit of even the least socially disadvantaged boys in the inner-city area does not seem to be a characteristic of working-class boys in the West Midlands conurbation as a whole. Phillips and Bannon (1968), using the Vernon test on a small, but well selected, representative sample of 11 year olds in Birmingham, showed that the mean reading level of boys with skilled manual working class fathers is very near the national average for their age, and even that of the combined group of boys with semi- and unskilled fathers is only about eight months lower (approximately one-third of a standard deviation). This is to be contrasted with our own results for the LSH and MSH boys, whose social-class origins overlap those two working-class groups: even allowing for the possible streaming bias, their likely mean reading ages are respectively about eighteen months and twenty-eight months behind their chronological age, corresponding to scaled scores of just over ninety and about eighty-five on the NFER test.

Our opinion that our results are not due to sampling bias is strengthened by further statistical analysis of our data, which shows that reading scores below expectation are a characteristic of both focus groups as well as the controls at both ages. Our colleague, Mr C.J. Phillips, has used his data on non-immigrant Midlands boys which form the basis of the local means in our histograms to calculate by multiple-regression techniques the predicted reading levels from age and score on the English Picture Vocabulary Test. The discrepancy between a boy's actual reading level and that predicted from his age and EPVT score yields a measure of retardation – the D score. D scores were converted to a conventional mean of fifty and standard deviation of ten for the Midlands samples. Scores above fifty indicate reading achievement greater than would be predicted from age and verbal ability; scores below fifty show reading achievement below expectation. Table 4.1 gives the mean D scores for our samples, kindly made available by Mr Phillips.

TABLE 4.1 Means and standard deviations of D scores

	6-7 years		10-11 years	
	Mean	SD	Mean	SD
Control LSH	40.8	6.6	43.6	12.5
Control MSH	46.1	6.5	41.4	9.7
Focus	41.5	6.2	44.6	12.6
Midlands boys	50.0	10.0	50.0	10.0

All three research samples at both ages have mean D scores much below the local mean of fifty – that is, they are well behind expectation on reading. Even the focus groups are considerably behind what is normally achieved by boys of their age and EPVT score in the Midlands. This demonstrates both that boys with their limited verbal ability can do better, and also that the result for the control samples may reflect a generally low reading standard

48 Chapter 4

in the inner-city areas. We shall discuss the reading results further in our final chapter.

We may now turn to a second consideration – the inter-group comparisons within our sample. At the side of each histogram in Figures 1-13 are the results of tests for these comparisons, using the data in Table 1 of Appendix C. It is clear that our expectation should be that the three groups will show a gradation downwards from LSH through MSH to focus, on the well-founded assumption that our social-handicap measure is associated with intellectual functioning as measured by conventional tests. Furthermore, because we have tried to arrange that the three groups at each age are clearly separated on the scale of social handicap, then it would be reasonable to predict that the separate pairs of groups (LSH and MSH, MSH and focus, LSH and focus) should all show significant differences. Thus we can test for the overall difference between the group means by simple analyses of variance, and for the difference between the pairs of means by another test resulting in the F statistic – the test for individual components of variation (Winer, 1962).

The prediction of overall differences between the means of the social-handicap levels is strongly supported in the case of every measure – except reading at 10-11 years. The biggest differences at both ages include the verbal measures, although the differences on non-verbal tests such as the Matrices, Draw-a-Man, and WISC Coding and Block Design are still substantial. There is only a slight indication that the more socially handicapped groups do relatively better on these latter tests. It will also be noticed from the histograms that the gradation down through the three groups each time is more or less a straight line, apart from the Southgate and Draw-a-Man tests at 6-7 years. The inter-group differences are less in the older age group, which could be due either to sampling effects or to the effects of schooling (if we are prepared to trust the comparability of the samples in other respects at the two ages, as seems reasonable).

Once again it is the reading test scores where the anomalous results occur. At 6-7 years the focus group show a much lower mean than the other two, which are very close to each other. Thus although there is a significant overall difference, it is comprised almost entirely of the difference between focus and both control groups. At the older age the overall F is not significant for either of the reading tests, and the LSH versus focus comparison for the Vernon test is the only significant difference emerging.

It seems that at least in the 10-11 year old age group, when the boys have been in education for over five years the less socially handicapped groups have failed to pull away very much from the most disadvantaged pupils. Their mean reading levels are a little higher than those of the focus boys, but the spread of scores in each group is very wide, so that the overlap between groups is considerable.

This pattern could be interpreted once again as indicating some difficulty in promoting the development of literacy even among the least socially disadvantaged. The LSH and MSH groups at 6-7 appear to have made a quicker start on reading, but it is not reflected in a further gain by 10-11. Our previous discussion on sampling factors and the consideration in chapter 12 of the possible reasons for the reading results is relevant here also.

The younger age group were given a larger battery of cognitive tests than the older boys, as will be seen from Appendix B. These tests do not have reliable external norms, but the indications are that the pattern of scores from them is very similar to that of the other cognitive tests: the LSH group are near average for their age, and the focus boys well below, the overall inter-group differences being highly significant. The relevant data for these tests appear in Appendix C. These additional results emphasize the all-round nature of the cognitive deficit in the focus boys, since the tests cover a wide spectrum, eg:

1. Piagetian items (Lunzer, 1970) examining the child's concepts of number, substance, and order using natural substances such as blocks, modelling clay and beads.
2. The basic use of a pencil in reproducing complex shapes, a skill known to be associated with later reading development.
3. The even more basic skill of the dot-joining task, where all that is required is accuracy and speed of vertical movement over the space of half an inch.
4. The reproduction of vocabulary and grammatical sequences from immediate memory.

The comparisons of the means of our samples in pairs shows that the LSH and MSH groups are generally, though by no means always, closer in ability than the MSH and focus groups. The LSH-MSH difference is usually only significant when the overall difference is very marked (i.e., when there is a steep slope down across the tops of the bars in the histograms). It might be argued from this that the focus boys have disadvantages exceeding some critical threshold, so that their mean scores show a sudden drop in the downward progression through the three groups. It is possible to examine this by using a test of the linearity of trend (the extent to which the difference between means can be accounted for by a straight line of best fit). We conducted such analyses and find that the straight-line assumption leaves no significant amount of variation for departure from linearity in any item (Phillips, Wilson, and Herbert, 1972; Herbert, 1972). There is therefore little evidence that the focus boys have disadvantages exceeding some critical threshold and producing a sudden 'bend' in the slope through the three means. Our conclusion should be a cautious one, since with only three groups and with a generally downward slope of means such as we usually have, departures from a straight line trend have to be very large in order to be significant. However, we feel fairly sure that the focus groups as a whole are not handicapped to some degree or in some way which makes them a completely distinct section of the local peer community.

BEHAVIOUR RATINGS

We now turn to the results from the ratings scales of classroom behaviour which were carried out by the boys' teachers. The basic question here is whether there are more behaviour difficulties with increasing social handicap. Common sense considerations might lead to this prediction, and some research evidence points that way too. For example, the National Child Development Study (Pringle, Butler, and Davie, 1966; Davie, Butler and Goldstein, 1972) found a sharp increase of 'unsettledness and maladjustment' associated with lower-social-class membership: on the Bristol Social Adjustment Guides (Stott, 1966) 58 per cent of social class V boys met the criterion score, compared with 41 per cent of social class III and only 32 per cent of social class I and II. West (1969) used a teacher completed scale developed from that of Douglas (1964) and combined it with ratings by psychiatric social workers to produce a scale of conduct. There is a highly significant association with 'social handicap' on West's definition of that term, which is similar to ours: of the severely social handicapped, 42 per cent were rated as 'badly behaved', but only $11\frac{1}{2}$ per cent of those with no social handicap (all subjects were boys and working class). Jackson (1972) found that among 'poorly adjusted' infant school children there was a significantly higher representation of children from materially and culturally poor homes (related to social handicap). In contrast, studies such as those of Mulligan (1964), Mitchell (1965), Chazan and Jackson (1971), Rutter, Tizard and Whitmore (1970), and others quoted by the latter authors, found little or no association between the behaviour problems of any type and social class.

We cannot account with certainty for the seeming conflict in these research findings. One possibility is that measures of social handicap such as West's tend to increase the likelihood of detecting children with behaviour problems, since they include more than just social class, for example family size which is associated with behaviour difficulties (Shepherd, Oppenheim and Mitchell, 1971). Our own belief is that instrument factors may also be at work. The National Child Development Study using the Bristol Social Adjustment Guide employed a simple social class division, yet found a marked gradient of behaviour difficulties. The total score used on that scale combines several types of problem, and therefore capitalizes on small trends in any single item or sub-scale. Furthermore, this instrument contains items which may be contaminated with educational or motivational factors to a greater degree than is usual: for example, items in the 'depression' group often seem to sample apathy and lack of persistence, those in the 'restlessness' section may sample poor motivation or dullness (for example, 'gives up easily', 'feckless, scatter-brain', 'can never stick at anything long'). Similarly, the scale used by West (op. cit.) includes items relating to school performance and attendance, as well as others concerning relationships and discipline. Gibson (1964, 1967) interprets this as a 'naughtiness' scale, yet clearly it samples much else besides behaviour problems as conventionally understood. The scales used by Jackson (1972) may also sample educationally related factors (Herbert, 1972).

Being aware of this possibility that educational and motivational factors may be more heavily represented in some scales than others, we decided to construct rating scales which attempted to differentiate between these factors and those of a more clinical type sampling difficult, withdrawn, or attention-seeking behaviour. With the aid of factor analysis, four sub-scales, each including several items, were devised at both ages:

A — anxious, worried, or tense behaviour v. relaxed, happy, unconcerned
B(I) — withdrawn, over-quiet behaviour
B(E) — acting-out, aggressive or impulsive behaviour
C — well-organized, persistent, and resourceful approach to school (we termed this 'competence')

In addition, the scales for the older age groups included items allocated to two further sub-scales:

D — demand for attention from (or perhaps dependency on) adults
E — sociability or extraversion towards children

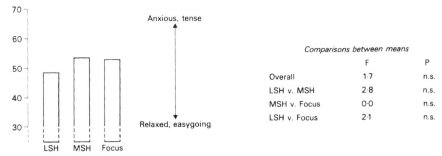

Figure 14 Behaviour Rating Subscale A: 6-7 year olds

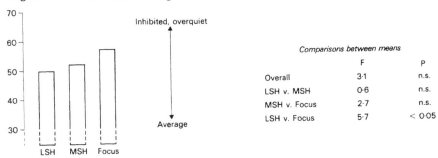

Figure 15 Behaviour Rating Subscale B(I): 6-7 year olds

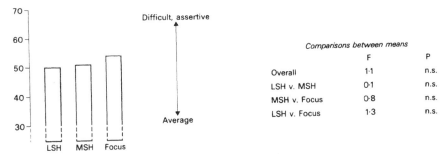

Figure 16 Behaviour Rating Subscale B(E): 6-7 year olds

52   Chapter 4

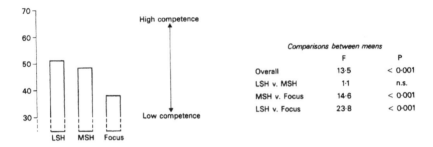

Figure 17   Behaviour rating subscale C: 6-7 year olds

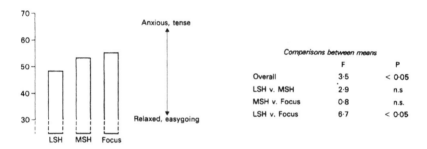

Figure 18   Behaviour rating subscale A: 10-11 year olds

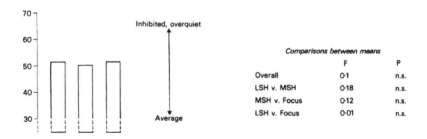

Figure 19   Behaviour rating subscale B(I): 10-11 year olds

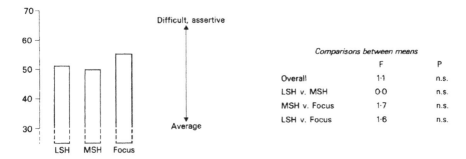

Figure 20 Behaviour rating subscale B(E): 10-11 year olds

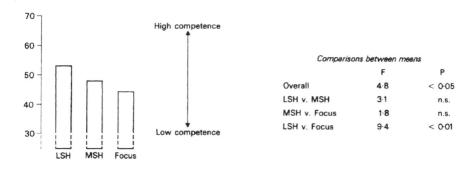

Figure 21 Behaviour rating subscale C: 10-11 year olds

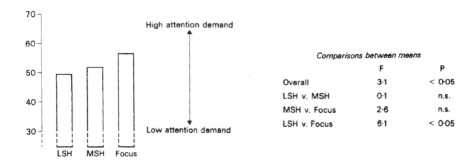

Figure 22 Behaviour rating subscale D: 10-11 year olds

54 Chapter 4

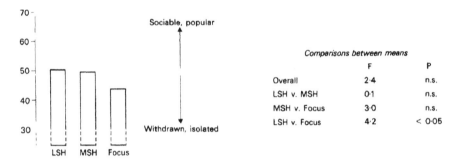

Figure 23 Behaviour rating subscale E: 10-11 year olds

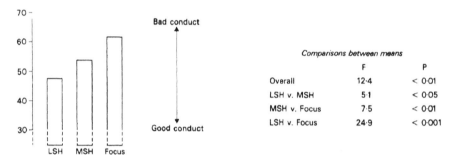

Figure 24 Conduct scale: 10-11 year olds

Appendix B gives an abbreviated description of our scales, together with references to more detailed accounts. The expectation was that the C (Competence) subscales would show the biggest differences between social handicap groups, since they included the kind of item which sampled educationally related behaviour as described above.

In addition, the 10-11 year olds were rated on the Conduct scale used by West (1969). Judging from his results, this scale would be likely to give a marked difference between our groups, and by examining its correlation with our own subscales we wished to see whether its description as a 'naughtiness' scale was justified.

We turn now to our data on inter-group comparisons within our sample. These are summarised in the histograms in Figures 14-17 and 18-24 for the 6-7 and 10-11 year olds respectively. The score base which is used is a mean of fifty and standard deviation of ten relating to the population from which the control boys were drawn. This mean is for most subscales probably around the average for an urban population. The data are given in raw score from Appendix C, Table C3.

In the younger age group the only significant overall difference between the groups is on the C subscale, where the gradation down through the means is almost as steep as that for the verbal ability tests.

Also like the ability tests the two control groups are closer together than the MSH and focus. The only other significant comparison is that between the least socially handicapped controls (LSH) and the focus boys on the B(I) subscale, where the latter appear more inhibited. In view of the non-significant overall comparison, this needs to be viewed with caution.

For the older boys there are more significant differences to report. The C subscale again shows the linear gradation towards lower Competence among the severely socially handicapped, although not so sharply as in the 6-7 year olds ($p < 0.05$ instead of <0.001). In addition, the A subscale shows a significant difference of about the same magnitude, increasing severity of social handicap being associated with increasing overt tension or worry. On the D subscale the significant gradient shows the more severely socially handicapped as on average more dependent or more demanding of adult attention. Finally, the Conduct Scale shows much the biggest between groups difference, in the direction of 'bad' conduct with increasing social handicap.

The C subscale has moderate to high correlations with reading, especially among the focus boys. The correlations are:

	Controls	Focus
6-7 years (Southgate)	0.37 ($p < 0.10$)	0.56 ($p < 0.01$)
10-11 years (NFER)	0.57 ($p < 0.001$)	0.72 ($p < 0.001$)

The correlations of cognitive ability measures with reading are no higher, and those of Competence and other cognitive measures are on average 0.15 to 0.20 lower, about the same as the average intercorrelation of the cognitive ability tests themselves.

We would not interpret this simply as due to the teachers basing their response to the Competence scale on the most overt evidence of cognitive functioning (reading). The questions contributing to the C subscales are given in Appendix B; they make no mention of reading as such, and it is an extremely round-about route to use reading as a basis for answering them. High correlations with reading probably arise from the fact that the Competence subscale measure what its name suggests - effective, self-organized, reliable, educationally motivated behaviour. In short, a teacher sees the competent child as knowing what being in school is about - not merely learning to read but getting on with the job, needing little supervision, organizing himself and so on. This adds up to a picture of the 'good pupil' (see Herbert, 1974b, for a further discussion of this concept in the literature). The child with severe social handicap is often not a good pupil in this sense.

How are we to interpret these findings? We feel that there is substantial support for our contention that the competence element is the most likely to yield significant mean differences between social groups. The two main 'clinical' problem subscales, B(I) and B(E), do not show significant overall differences at either age. Thus the most consistent and pervasive tendency is for increasing social handicap to be associated with low competence in the sense of poor concentration or perseverance, lack of self-reliance in the classroom, and low reliability. The results for the A and D scales at 10-11 years suggest that there is some tendency for the more socially handicapped to show manifest signs of stress in the school situation by being more overtly anxious (for example, by unease or

56 Chapter 4

by becoming flustered), and by seeking adult attention. At the
younger age, if the B(I) result is to be credited, the most severely
socially handicapped actually tend more often to do the opposite -
to withdraw or to be inhibited.

The highly significant inter-group difference on the Conduct
scale can now be seen as at least in part a reflection of the competence differences between the groups rather than an indication
that severe social handicap simply produces more 'naughtiness'

TABLE 4.2 Correlations[1] between behaviour rating scales

(a) 10-11 year olds

	Conduct	E	D	C	B(E)	B(I)
A	29**	-36***	-07	-32**	-16	37***
B(I)	-18	-52***	-43***	-01	-57***	
B(E)	58***	14	23*	-28**	-	
C	-73***	16	-05	-		
D	11	40	-			
E	-23*	-				

(b) 6-7 year olds

	C	B(E)	B(I)
A	-53***	10	33**
B(I)	-22*	-55***	-
B(E)	-17	-	

1 Mean product moment correlations of the three social-handicap
groups, calculated via 'z' transformation. Decimal points omitted.
Significance levels (two tailed) *** $p < 0.001$ ** $p < 0.01$ * $p < 0.05$.

Table 4.2 gives the intercorrelations* between the subscales at the
two ages. The Conduct scale has a very high correlation (-0.73) with

* These are mean product-moment correlations, each one being the average of the separate correlations from the three social-handicap
groups, after ascertaining that the three were not significantly different. Fisher's z transformation was used. A correlation coefficient
can vary from +1.0 to -1.0: the former indicates that the two variables
go up or down exactly in step with each other; the latter indicates
that high scores on one are associated with equally low scores on the
other. Values falling between these two extremes show increasing degrees of uncertainty as to what the score on one variable will be when
the other is known. The uncertainty increases with smaller samples,
but in general a coefficient of 0.5 denotes a moderate degree of association. A value of 0.0 always means no association, and a non-significant correlation is assumed not to be different from 0.0.

the C subscale at 10-11 years, which is much higher than its association with any other subscale. The correlation with B(E) is moderately high at 0.58, which shows that as well as sampling competence it also samples difficult behaviour to some extent. Gibson's description of it as a 'naughtiness' scale is therefore justified only in part, since the biggest element is clearly competence, and judging by our results it is this which contributes heavily towards the Conduct scale's discrimination between social-handicap groups. The other correlations between subscales are what one might expect and little comment is required. It should be remembered that the B(E) and B(I) subscales are composed of the opposite poles of the same items, and therefore have skewed distributions which tend to lower their correlations with other scales. The moderately high negative correlation between the A and C subscales at both ages should be noted: it seems that high anxiety as measured by A is linked with low competence, an indication that the A scale may pick up signs of tension which are related to a child being out of his depth in school.

So far our consideration of the behaviour-rating results has used parametric statistics: those relying on the distribution of scores within each group as a whole and assuming that this distribution centres around a mean which is the mid-point of a reasonably bell-shaped distribution. Significant differences between groups on such assumptions imply that the groups have distributions centred on means which are sufficiently separate to make the overlap of the distributions small enough to be beyond chance levels. Thus parametric statistics tell us something about the characteristics of a group as a whole (expressed in terms of its mean score).

The behaviour-rating scales all show a reasonably bell-shaped distribution of scores except for the B(E) and B(I) scales, which were constructed in such a way as to put the bulk of children near one extreme, with a long tail of high scorers at the other extreme. In fact, both these latter scales sometimes have a secondary bulge of very high-scoring children super-imposed on this tail, so the distribution is not at all bell-shaped. Generally speaking, parametric statistics for group comparisons are known to be 'robust' when assumptions of normal distribution are violated, so it is not likely that our lack of significant inter-group differences for the B(E) and B(I) scales can be ascribed to their odd distributions. We take the non-significant, overall differences on these scales to imply that most of the severely socially handicapped are not distinguishable from their peers in the same school as regards the dimensions of difficult or of inhibited behaviour. However, the peculiar distribution of B(E) and B(I) scores does raise an additional question: is there a large proportion of focus boys among the high scorers on these two scales? This question can be examined by using distribution-free statistical tests where the question is whether in one group there are more subjects above a certain criterion level of score than in a comparison group.

Since we are interested in adjustment in the clinical sense, rather than just the B(I) and B(E) results, we also looked at the A subscale in a similar way, since between them these three seem likely to detect most children with adjustment problems in school. For each of these scales we adopted a criterion score of sixty-five or over as indicating quite severe problems – i.e., one and a half standard deviations above the control population mean. When we counted the numbers of controls and focus boys having such scores, the results were as shown in Table 4.3.

TABLE 4.3 Boys having scores of 65 or over on comparison of focus boys with combined control groups

Subscale	Controls		Focus		χ^2 df=1	p
	N	%	N	%		
A	12	10.3	11	19.3	2.00	ns
B(I)	11	9.4	12	21.1	3.58	ns
B(E)	11	9.4	10	17.5	1.69	ns
Any one of above	28	23.9	22	38.6	3.34	ns
Total groups	117	100.0	57	100.0	–	–

The combination of the age groups gives larger numbers in the cells, and is justified because their contributions are evenly balanced. None of the comparisons between the numbers of controls and focus boys scoring over sixty-five is quite significant, but it is noteworthy that there are consistently almost twice as many focus boys above the criterion score. This indicates that the incidence of severe behaviour difficulties is higher among the severely socially handicapped, and leads us on to a further statistical test: are there fewer well-adjusted boys in the focus samples? If we adopt as the criterion of good adjustment that a boy should score no higher than fifty on any of the three subscales A, B(I) or B(E), the results are as in Table 4.4.

TABLE 4.4 Boys having no score above fifty on subscales A, B(I), and B(E) (= 'well adjusted'). Comparison of focus boys and combined control groups

	Controls		Focus	
	N	%	N	%
Well adjusted	25	21.4	4	7.0
Remainder	92	78.6	53	93.0
Total	117	100.0	57	100.0

chi-squared = 4.70   df = 1 $p < 0.05$

Once more the situation is similar within each age group, so their combined results are presented. There are significantly fewer well-adjusted focus boys. This complements the previous findings, and enables us to be fairly sure that among the severely socially handicapped there is a larger minority with adjustment problems. We would again stress, however, that our parametric statistics indicate that most of the focus boys had behaviour indistinguishable from the majority of their peers in school.

C - SELF-REPORT SCALES

We will now report the results from questionnaires completed by the 10-11 year olds only. This will be done in a selective way for the sake of clarity and brevity. Detailed discussion will be found in Phillips, Wilson, and Herbert (1972) and Herbert (1972).

There were two self-report instruments. One was the New Junior Maudsley Inventory (Furneaux and Gibson, 1966), which is a published questionnaire for measuring the well-established dimensions of Extraversion (E) and Neuroticism or Instability (N). All items are statements about personal preferences or attitudes, and the subject is required to say whether he is the 'same' or 'different'. Extraversion items ask about the subject's behaviour in groups or in situations where he has the choice between a quiet self-involved activity and a more active or sociable one. The Neuroticism items concern the subject's susceptibility to worry, irritation, and feelings of tension. In addition, the NJMI incorporates a 'Lie' (L) scale, where positive response involves assenting to very self-approbatory statements.

The other self-report questionnaire was a Self-Concept test adapted from Wilkie (1962). The items appear in Appendix B. The format was similar to that of the NJMI: statements such as 'I usually do better than other boys in class', to which the subject had to reply 'same' or 'different'. The items were of two types - first, Evaluation (twenty-four items) where the subject assesses himself in terms of school performance, financial circumstances, peer-group acceptance, etc., and second, Motivation (eight items) where the items refer to aspirations such as 'I want to come top and beat others'. A second administration was used for the Ideal Self score, where the boys were required to say whether they would like to be the same as or different from the boy in the statement. The items were scored in such a way that a high total indicates a favourable view of the Self, and a good report about motivation. The discrepancy between Self and Ideal Self was calculated as indicating the degree of self-acceptance or self-dissatisfaction. The Ideal Self and discrepancy scores were not used in the research because it was found that one of the authors (GWH), who administered tests to about two-thirds of the sample, produced significantly higher Ideal Self scores in his subjects compared with the other psychologists contributing to the project.

Both the self-report questionnaires were originally designed for group administration. In the present research they were administered individually, and the psychologist read the items to those boys who were unable to cope with the reading themselves. This adds a component which complicates interpretation, since it may be implied that individual administration increases the likelihood of self-consciousness.

Table 4.5 gives the results of the NJMI in raw-scores form, together with comparison data from McAllister and Marshall (1969), who tested a sample of 250 11 year olds, an age range nearer that of our own boys than the 10 or 12 year olds in the NJMI manual.

TABLE 4.5 Data from the research samples and a normative sample on the New Junior Maudsley Inventory: 10-11 year olds

McAllister and Marshall (1969)		Extraversion		Neuroticism		Lie	
		Mean	SD	Mean	SD	Mean	SD
Boys and girls		12.0	2.9	8.9	3.3	-	-
Boys only*		-	-	-	-	7.1	3.4
Research samples	LSH	12.0	3.0	8.0	3.4	10.6	4.0
	MSH	12.8	2.1	8.0	3.2	9.1	3.9
	Focus	11.4	2.4	8.8	2.5	9.8	3.4

* Sex differences on Lie Scale only.

All three groups score close to national norms on the E and N scales, but much higher on the Lie scale. Individual administration may account for this high degree of defensiveness or unrealistic self-approbation, but it is according to expectation that L scores are high in the manual working class (Eysenck and Eysenck, 1963). Gibson (1969) reports that a socially handicapped group within a working-class sample had higher L scale scores than their peers in the neighbourhood. In our sample the inter-group differences on this scale are not significant, nor are they for the E or N scales.

On the Self-Concept test to external comparison data are available. As can be seen from Table 4.6 there are no significant inter-group differences in our sample.

TABLE 4.6 Means and standard deviation of self-concept tests: 10-11 year olds

	Control Low SH		Control Moderate SH		Focus	
	Mean	SD	Mean	SD	Mean	SD
Self-concept Evaluation	12.7	4.0	11.5	3.2	11.7	3.6
Self-concept motivation	4.4	1.6	4.6	1.3	4.8	1.5

The correlations involving the self-report measures give some hints as to what is happening as it were behind these results. The more definite findings are:

(i) The control boys show much more significant correlations within a subset of the self-report measures. Table 4.7 illustrates this.

TABLE 4.7 Average correlations in the control groups compared with focus group for a selected self-report measure

Items correlated	Mean correlations* from control groups	Correlations from focus group
NJMI lie with self-concept evaluation	−0.53 ($p < 0.001$)	−0.04 (ns)
NJMI lie with NJMI neuroticism	0.62 ($p < 0.001$)	0.19 (ns)
NJMI neuroticism with self-concept evaluation	−0.45 ($p < 0.001$)	0.19 (ns)

* Here and in subsequent tables where mean control correlations are given, it was first ascertained that the correlations from the two control groups were not significantly different.

The common strand running through these subscales is that each can be considered to have a 'good' and a 'bad' pole. Compared with the focus boys, the controls seem more able to place themselves consistently at one end or the other. Since there is no significant difference between the means of the groups on any of these scales, the controls as a group do not appear to place themselves uniformly at one end of the continuum.

(ii) The control boys also show more consistent relationships between their own ratings of themselves and their teachers' ratings, again at least as regards one subset of measures. The relevant correlations are given in Table 4.8.

TABLE 4.8 Average correlations in control groups compared with focus group on New Junior Maudsley E Subscale and Selected Behaviour Rating Scales

Items correlated: NJMI 'E' and:	Mean correlation from control groups	Correlation from focus group
Behav. Rating B(I)	−0.35 ($p < 0.01$)	−0.23 (ns)
Behav. Rating B(E)	0.33 ($p < 0.01$)	−0.05 (ns)
Behav. Rating D	0.39 ($p < 0.01$)	−0.02 (ns)
Behav. Rating E	0.26 ($p < 0.05$)	0.39 ($p < 0.05$)

Here the common strand is the extraversion component: all correlations involve NJMI E and a behaviour rating subscale which taps the introversion-extraversion dimension. Here, too, the control boys seem more able to present a picture of themselves, this time in terms of the extraversion continuum. The picture also seems to have some validity as judged by the external criterion of teachers' ratings.

(iii) The control boys show more consistent association between the Self-Concept Motivation scale and measures likely to reflect academic achievement or school-orientation. Table 4.9 shows this in the correlations with the two reading tests and the behaviour-rating Competence subscale.

TABLE 4.9 Average correlations in control groups compared with focus group of Self-Concept Motivation Scale and selected other measures

Items Correlated: Self-Concept Motivation and:	Mean correlation from control groups	Correlation from focus group
Vernon Reading	0.31 ($p < 0.01$)	0.06 (ns)
NFER Reading	0.31 ($p < 0.01$)	0.00 (ns)
Behav. Rating C	0.35 ($p < 0.01$)	0.07 (ns)

Thus it seems that the control boys have a clearer and more consistent self-image in terms of motivation, just as they do in terms of the good-bad and extraversion continua. We would favour the interpretation that the control boys are more aware of the various normative yardsticks which society uses, and can measure themselves at least roughly on such criteria. In doing so they may, as with the NJMI Lie scale, present an overgeneralized picture, but this at least shows that they are aware of the evaluative criteria. The focus boys, in contrast, did not seem able to perform such self-evaluative judgments with consistency. Perhaps this is because of the questionnaire format in which they were required to operate: being duller, on average, than the control boys they became confused and resorted to guessing - but in a way all self-assessment has the same format, in which the subject asks whether he is the same as or different from a given standard.

SUMMARY

A full summary of the psychological test results appears in chapter 12. In brief, the cognitive tests show the wide range of ability to be found in an inner-city area; they also show considerable association with social-handicap level, a relationship which holds good for many different types of ability. In reading, even the least socially handicapped group does not do very well, so that by 10-11 years the three social-handicap levels are not significantly different from each other.

In the behaviour rating scales the salient results are the significant tendency at both ages for increasing social handicap to be associated with lower competence in the sense of self-organized well-motivated conduct in school. Further, at 10-11 years there was a significant social-handicap difference on the scale measuring anxious behaviour, with severe social handicap being linked with higher scores on this dimension, and also there was a similar effect on a subscale measuring dependent or attention-demanding approaches to adults in school. There was only a slight tendency for severe social handicap to be associated with problem behaviour of an aggressive or overquiet sort, although combining these and the 'anxiety' subscale into a category of behaviour difficulties showed that there was a non-significant trend for more focus boys (severe social handicap) to have extremely adverse scores, and a significant tendency for the control boys (low or moderate social handicap) to have a higher proportion of well adjusted boys.

On self-report scales there were no significant differences between the social-handicap groups, but there were indications from correlational evidence that the control boys were more able to rate themselves consistently and in line with external criteria.

Chapter 5

THE PLAY-GROUPS

Play-groups for the 3 and 4 year old children were conducted on two mornings each week during the autumn and spring terms of the two years' field work. Much of the first term was devoted to developing techniques of observation and recording, and to adapting traditional nursery school practices to the needs of a group of young children quite unaccustomed to the environment of a well-equipped play room. Although most of the families sent their preschool child or children to the play-groups, attendances were irregular, so that observational records are based on thirty-one children who had attended often enough to make accumulation of these possible. Towards the end of the second year a control group of children was introduced of the same age range and sex composition. These children, sixteen in all, attended a play-group in a mixed residential area of the city close to the university; they came from working-class families mainly in the skilled category; also their families were much smaller than those of the main sample.

   The children from the main sample of families were collected (by HW) by minibus, and after the morning's session returned to their homes. In some cases the children needed much personal attention; sometimes they had to be got out of bed and dressed, more often help was needed in locating a missing shoe, or in supplying essential clothing like knickers or a warm pullover which was always carried on the bus. Attendance was irregular in spite of transport on account of frequent ill-health or occasional unwillingness to leave home. In some families the children were always ready when called for, and often waiting at the door. On arrival the children gathered in a domestic setting in the Play Centre and were given drinks and biscuits which helped to relax them; there were always some children who had missed breakfast at home.

   Two rooms were used by the children. A large play room, equipped with wheeled and constructional toys, wendy house, sandbox and wet-play area, dressing-up clothes, etc., was connected with an observation room through a one-way vision screen and six *in situ* microphones. The children's workshop was used for painting, cutting-out, pasting, and clay-modelling. Movement between these two rooms was uninhibited and left to the children. Towards the end of the morning the children assembled in the play room for a short story. The children wore overalls each of which had a distinctive symbol on front and back, such as half-moon, scissors, circle, etc., by which they were identified by the observers.

The children were given individual psychological tests, using both verbal and non-verbal measures; they were observed at play, and recordings were made of spontaneous speech. The latter two measures were repeated on the group of controls for comparative purposes. Neither tests nor recordings took place until the child was well settled and familiar with his surroundings. Children in the play room were observed through the one-way vision screen, but those in the workshop had to be watched by observers standing in the doorway or sitting in the room. The observers were mainly postgraduate, mature students. As is usual in observational studies the children largely ignored them when it was apparent that they would not interact. One of them was detailed each session to make a written recording of language and spent a set amount of time in each room.

THE PSYCHOLOGICAL TESTS

The first year's group of children was used in order to try out various measures of testing. This process could not start until the children were settled in the play situation and were familiar with the examiner (GWH) who made himself known to them by helping with biscuits and drinks, and by being present in the play room for some of the time.

Most children had adapted to the situation quite happily after two or three sessions, but many remained reluctant to go into another room with the examiner. Various strategies were adopted to overcome this; sometimes an older sibling, who had accompanied the child to the play session, would bring him into the test room; sometimes a bolder child, already tested, would accompany the less forthcoming ones. As a last resort, the examiner moved into the play room and attempted to test the child in a corner, competing with more attractive toys. Although experienced with testing young children, the examiner found this group to present more difficulties than any other in his acquaintance. Some were reticent or fearful, some very distractable and difficult to keep in the test room. The pervading impression of the group is one of bewildered, 'lost' children, who did not know what to expect, who continued to be worried, and who seemed unable to relax; an impression which was reinforced by their very poor clothing, unkempt appearance and smell. Not all the children were like this, but the sturdy, self-possessed ones stood out as exceptions, although their behaviour was really no different from average children of their age. In addition to the problem of adjustment to the test situation several children put in a very irregular attendance; others, who had attended regularly, were admitted to school before they could be tested. The result was a considerable reduction in the number of children on whom tests could be conducted. In the first year a total of twenty-seven children made some attendance, but tests could be conducted on only fourteen of these. In the second year thirty-one children attended, but only sixteen could be tested. These figures make it difficult to evaluate the representative nature of the sample obtained. It can be stated with some certainty, however, that the sample does not consist of the more adjusted, settled children only, as regularity of attendance was not related to the child's adjustment; and some children co-operated with the help of an older sibling who accompanied them.

In the first year the Reynell Developmental Language Scales (Reynell, 1969) were used. These comprise an Expression Scale requiring verbal responses from the child in naming objects and pictures and in describing the activities shown in pictures. They also comprise a Comprehension Scale which needs no overt verbal response, but tests the child's understanding of instructions of varied complexity as shown in the way he manipulates a set of toys. This test is fairly long, necessitating a second session for the non-verbal measures. In the second year a shorter linguistic measure was used, the Peabody Picture Vocabulary Test, Preschool Version (Dunn, 1959). This test has greater discrimination among the less able young children. Administration of this test was sufficiently quick for the non-verbal measures to be given in the same sittings in most cases. It uses one-word stimuli from the examiner and is thus a simpler test of comprehended vocabulary with less interference from factors like willingness to co-operate, or to listen.

The non-verbal measures chosen were all from the Merrill Palmer Scale (Stutsman, 1931). This is a standardized collection of children's toys such as bricks, nested cubes and form boards, which most pre-school children love to do. The norms are rather out of date but suffice for a rough classification. The complete test was too long and would have taxed the limited co-operation of many children in the group; for that reason a number of sub-tests were chosen. These consisted of building a tower of one-inch cubes, constructing a three-cube pyramid when shown a model, putting sixteen cubes back in a flat box, completing the ten-piece Seguin formboard, and putting together a wooden mannikin. These activities represent an age-range from 18 months to about 6 years, and from their performance the children could be assigned to one of nine 'mental age' categories, each covering a range of 6 months. Age levels were usually covered by more than one subtest. A child was assigned to the best mental-age category he achieved. This procedure is undeniably crude, but our experience of these subtests on other children encourages us to believe that they do not give misleading results. Since both tests, the Reynell used in the first year and the Peabody used in the second year, give 'mental age' scores, the children's achievements could be converted to the simple nine-category system.

Table 5.1 gives the results of the psychological tests for both years. The language measures consistently show a tendency for mental age levels to fall below the chronological age. The similarity of the Reynell and the Peabody test distributions is striking and indicates that this linguistic retardation on the Reynell test in the first year group was not an artefact of that test's demands on the child's co-operation. The non-verbal results in both years show a bimodal distribution: some children did quite well, others rather badly, with very few in the intermediate position. The Merrill-Palmer

TABLE 5.1 Pre-school play-groups psychological test results. Numbers of children in stated categories of mental and chronological age

	1^6-1^{11}	2^0-2^5	2^6-3^{11}	3^0-3^5	3^6-3^{11}	4^0-4^5	4^6-4^{11}	5^0-5^5	5^6-5^{11}
CA (first year)	–	–	–	3	6	5	–	–	–
CA (second year)	–	–	2	2	5	3	4	–	–
CA (both years)	–	–	2	5	11	8	4	–	–
RDLS expression* (first year)	1	2	6	1	2	–	–	–	–
RDLS comprehension** (first year)	1	4	5	1	2	–	–	–	–
PPVT (second year)	2	5	6	1	2	–	–	–	–
RDLS comprehension and PPVT	3	9	11	2	4	–	–	–	–
Non-verbal (first year)	1	–	5	4	–	4	–	–	–
Non-verbal (second year)	1	–	2	–	–	6	5	1	1
Non-verbal (both years)	2	–	7	4	–	10	5	1	1

* two children omitted } due to co-operation problems.
** one child omitted

Scale in general tends to show this characteristic. Work on its content (Stott and Ball, 1965) has demonstrated that the abilities sampled by this scale are rather different from those sampled by traditional non-verbal reasoning tasks at later ages: the Merrill-Palmer contains many more items tapping perceptual-motor or manipulative ability.

We would interpret success on this scale as indicating an intact nervous system and at least a basic acquaintance with geometrical form and manipulative tasks. These latter skills could be acquired even in a relatively deprived background and are not necessarily predictive of good non-verbal performance later on. Failure on the Merrill-Palmer items could be associated with brain damage, and this may be the case in one or two children in our group. However, a more likely explanation in most cases would be a general low ability combined with reticence.

In the first year, when a whole sitting was devoted to non-verbal testing, the children were given the Merrill-Palmer design copying items. Despite the fact that five of them reached the level of a $4-4\frac{1}{2}$ year old on the Seguin formboard, none reached even the $3\frac{1}{2}-4$ year level in copying a cross. The best any child could do was to copy a circle, which represents the 3 year old level. It appears that the children are unaccustomed to the use of pencil, and their ability in copying geometrical shapes seems less well-developed than the perception of shape itself. This is not an unusual finding with the Merrill-Palmer Scale in clinical use, perhaps because so many pre-school referrals are in the socially deprived groups.

Our impressions of these results are twofold. First, as expected, they show a very slow start to language development made in these families. Second, in many cases perceptual and manipulative skills were relatively intact. Thirteen children (43 per cent) had a non-verbal mental age equal to or greater than chronological age, in other words, they were of average ability or above. This surprising finding needs to be viewed against the rather poor results on non-verbal tasks in the 6 to 10 year old boys from the same families. It seems that the perceptual and manipulative skills acquired by these children in their pre-school years are not those needed for success later on.

BEHAVIOUR: SOCIAL INTERACTION AND MATURITY OF PLAY

Observation techniques

The most commonly-used method is the time-sampling checklist where an observer, watching a single child, makes marks in cells formed by rows for categories of behaviour, usually less than fifteen, and columns for time intervals, usually of ten or fifteen seconds. Variations on this method, described by Hutt and Hutt (1970), include the use of symbols for behaviour categories, mechanical recording, and speaking a code into a tape-recorder. Such methods have the advantage of quantitative data, by frequency of precisely defined behaviours per unit time.

When investigators have sought to describe behaviour, rather than to record the occurrence of previously-selected categories, they have usually employed some form of narrative recording, by written or spoken means. Narratives may have no prescribed time limits, or are based on relatively long time-samples. Examples are the work of Lunzer (1955), Wright (1967) and Patterson, Littman and Bricker (1957). The first two authors mentioned use two steps to reduce the narrative to scorable form, first dividing it into 'behaviour units', and then scoring these by rating scales. Patterson applied a checklist directly to the narratives, having only types of aggressive behaviour to analyse. In addition to these references, the work edited by Blurton-Jones (1972) provides many examples of methodological problems encountered in this study.

The eventual form of recording adopted in the present study was the narrative, scored later with a checklist. This choice was made after a lengthy pilot stage, when the time-sampling method was attempted. A list of behaviour categories was devised for this, based mainly on the pre-school social behaviour items used by Smith and Connolly (1972), made available by the authors in a preliminary version.

The Smith and Connolly items were reduced in number, and supplemented by further items sampling play in terms of the material employed and the ways in which it was used. This checklist was over twenty-five items long, and observers found it a strain to bear all the categories in mind. This problem could no doubt have been solved if the observers had been able to devote more time to learning the techniques, but they were mostly teachers on advanced courses, who could only come once a week for an hour or so. The narrative method of recording obviated this difficulty, and also it greatly reduced the observer training time, making more sessions usable as a source of data for the study. Another advantage of narrative recordings was that the choice of categories for analysis could be left until after the fieldwork, and a longer list could be used. The final list could, therefore, be decided when the children's behaviour repertoire was more fully known.

The disadvantages of narratives are the obviously selective nature of the primary recording, the time required to produce readable protocols, and the indirect nature of the final data. Despite these drawbacks, narrative recording was plainly the correct choice in the present research.

The number of children to be observed, up to ten in a group, and the number of observers, up to six at a time, precluded the regular use of tape-recording. Narratives were written by hand. Along the lines of Lunzer's work, the observers wrote for one minute on the play of one child, after which a buzzer sounded for them to proceed to the next one-minute section of the sheet they were using. They wrote on one child for five one-minute intervals, and then changed to another child, on a rota basis.

The time-sampling plan provided for the observation of the child by different observers at different times within one session, and his final collection of records could be selected to give a spread of observations across and within sessions. It was found that few activities extended beyond five minutes in the focus groups. Lunzer used two observers of one child, recording simultaneously, who later produced an agreed account. It is doubtful whether much increase in accuracy arises from this. Wright (1967) and Patterson (1957) did not use independent simultaneous narratives, except in training observers, as we did.

The result of this method of recording is a series of sheets, each containing five minutes' observations on one child. These were scored by checklists. On the checklists each minute was notionally divided into three intervals so that a behaviour category could receive fifteen marks in one five-minute sample.

The observation and scoring methods, and the checklist content, were subjected to a long period of development which will not be presented in detail here. The process was designed to ensure that the checklist items were those with a reasonable rate of occurrence, that raters using the checklists had close agreement, and that the time samples included for each child gave a representative cross-section of his activity.

Checklists of behaviour categories

Full specifications of the two checklists, Maturity of Play and Social Interaction, are given by Herbert (1974a). Briefly, the checklist items were as follows:

List I 'Maturity of Play'

A *Miscellaneous*
   Static, doing nothing
   Takes an object
   Puts down, drops it
   Holds, carries it
   Walks, runs
   Sits, kneels at a task
   Wears dressing up clothes
B *Speech in Play*
C *Low-level exploratory activity*
   Examines an object, or substance in terms of basic physical
      properties
   Clambers, jumps
D *Instrumental Activity*
   Uses small (for example, cup, gun, toy car, used by itself and
      not part of a larger role sequence)
   Uses large (an object above table-toy size is used for a mechanical function - for example, truck, see-saw - but is not
      used to construct anything)
E *Throwing*
   Throws ball or other object
F *Imaginative, creative and symbolic activity*
   Arranges objects in a heap, line, etc.
   Role Play
   Constructional Play, making something by fitting or joining
G *Specific creative materials*
   Pencil (including crayons and chalk)
   Scissors
   Plasticine

List II 'Social Interaction'

A *Interaction types*
   By Self

Parallel play (with same material as another child nearby, but
   not *with* that child)
Group Play (in group where members each contribute)
Non-play with other(s) (interaction with other child but not in
   play)
With older child (for example, an accompanying older sibling)
With adult

B *Miscellaneous*
  Speech: S = to self   These subdivisions were also
           C = to another child   used for other categories
           A = to adult   as appropriate
  Shout
  Crying
  Smile
  Looks at others
  Looks around (no obvious fixation)

C *Non-aggressive behaviour*
  Gives, holds out an object to another
  Non-aggressive passive (doesn't assert himself against approach
    by another, for example, obeys, retreats)

D *Negative interactions*
  Hit, push, pull
  Aggressive, existive

E *Adult activities*
  Adult helps, directs
  Adult checks, directs

Results from behaviour observations

Maturity of play will be discussed first. Here a clear pattern
emerges. As shown by a test for differences between means, the
focus group has a higher incidence of the following behaviours:
takes, puts down, examines, uses small object, and throws. Thus
they seem to spend more time on what may be termed basic manipulative exploration of the play environment. Despite the greater
investigation by handling and using toys, they do not produce any
more play of a more complex nature than do the Controls: for example, the groups spent about equal time on constructional and role
play, or on using paint brush, drawing implements, scissors, and
plasticine clay.
  In the checklists we tested each item for the significance of
the difference between the spread (or variance) of scores in the
focus and control groups. In the Maturity of Play checklist, of
eighteen such tests which were computed, no less than ten showed a
significantly wider spread of scores in the control group, often
twice as wide. Interpretation of this is difficult. Perhaps it
is due in part to the fact that some of the control subjects had
fewer total samples of behaviour. In such subjects an occurrence
of a behaviour would carry more weight in percentage terms. At the
opposite extreme in these same subjects with short totals of sampled
time, there would be more chance that certain behaviours would not
be recorded at all, thus giving them zero scores. These *two pheno-
mena* would yield increased variances. Therefore the differences

between variances may be an artefact.

An argument against this is that the Social Interaction checklist shows far fewer significant differences between variances (three out of fourteen computed, two of them at the 0.05 level only). Thus the focus group may have a narrower spread of scores more often in the cognitive aspects of play than in the social aspects. If this is a psychologically real phenomenon, it indicates the homogeneity of the play patterns in the focus families, contrasted with greater individual differences in experience and functioning among the controls.

Turning to the differences between means in the Social Interaction checklist, a clear pattern again emerges. The focus group spent more time alone, looked about them more often, looked at adults more often, initiated more contacts with adults, and were more frequently helped by adults. There is also a nonsignificant tendency for them to spend more time in parallel play (engaging in same or similar activity as others nearby but not co-operating) and less time on group play (co-operative activity where each has a complementary role). Thus the total picture is that the focus children were less integrated with their peers, spending almost half their time alone, and oriented themselves more towards the adults. The adults responded to this by giving more help to the focus children.

These results fit our impression of the play as it occurred. The controls were a more cohesive group, more independent of adults. Although the situation was new to them and recording commenced immediately, they clearly settled in very quickly and took over the premises without hesitation. The fact that they were drawn from an ongoing play-group in the locality of the university may account for this. In contrast, the focus children were much less at home even after a settling-in period, and were more inhibited about social contacts with the peer group.

We would summarize the behaviour observations in the play-group as highlighting the likely conduct of children from socially handicapped families when they are brought into an educational setting for the first time. This is totally new to them and their reaction is to explore tentatively, restricting their interest to small manipulable objects. These objects are frequently those with which they are familiar - guns and cowboy outfits, toy cars, adult clothes, pots and pans, water and containers, hammers and so on. Often these are merely carried, or briefly held and put down. In social contacts they are inhibited, but learn to approach the adult staff. The contrast with the comparison group is marked, but this group differed in more than one crucial respect - it was less socially handicapped and it was from a well-run nursery setting. This means that we cannot interpret our results as showing the single effect of social handicap or of nursery experience. What they indicate is the huge task which confronts any nursery-school staff when they take in a child from a socially handicapped family.

Chapter 5

LANGUAGE OBSERVATIONS IN THE PLAY-GROUPS

Various methods of recording the spontaneous speech of the children were tried. Although *in situ* selective microphones were in use in the children's activity area in the play room, it was found that background noise made speech inaudible for much of the time in tape recordings. Therefore it was necessary to resort to written recordings of speech. The disadvantage of this method is that the writing speed of the observer limits the amount that can be recorded when there is a lot of talk going on or when the utterances are longer and more complex. The result is a more adequate recording of the speech of the deprived children: observers complained at times of not being able to write down all that was said by the controls. Differences between the two groups tend on that account to be underestimated.

The children's spontaneous language was recorded in two situations: in the large play room and in the workshop equipped for painting and model-making. During the half-day play session an observer, on a rota basis, spent half an hour at set times in each room. The plan aimed to give every child an equal chance of being recorded; we believe it achieved this reasonably well. The recorder sat unobtrusively in the room as near to the children as possible. Absorbed in his task, and avoiding eye contact with the children, he seemed unnoticed by them most of the time.

Utterances made by every child were taken down verbatim; the identity of the child was noted by the symbol on his overall. Notes were made of the type of activity engaged in by the child when he was speaking, and symbols were used to identify whom he addressed. The persons with whom communication occurred are divided into 'other children' and 'adult'. A third classification, 'speech to self' was used so rarely that it is not included in the analysis. The original recordings were transcribed, collecting all utterances from individual children to form a record covering all their utterances.

Many measures of analysing the recordings were considered; for example counting the number of morphemes, scoring the uncommonness of words in a vocabulary study, or using a scale of sentence structures. The measures finally adopted were the only realistic ones within the research resources, although very time-consuming, and they could not have been realized without the help of a number of volunteers. The three measures we adopted were:
1 average length of utterances, obtained from the total number of words divided by the total number of utterances for each child;
2 average number of words per attendance, which is the total number of words spoken divided by the number of occasions when the child was present while language recording was taking place;
3 average number of utterances per attendance, which is the total number of utterances treated in the same manner as the preceding item related to words.
In general there was no difficulty in counting the number of words; 'doesn't', 'isn't', etc., were counted as two words. An utterance is a sequence of connected speech, including merely a single word, not broken by a significant pause or an utterance from another person. Briefing sessions held from time to time with the observers ensured conformity of practice.

The first of these measures is the most important indicator from the point of view of the development of the child, since it reflects the complexity of syntactical structure rather than the sheer amount of talking. The longer the utterance, the more complex it is likely to be. But it is not a sufficient measure of language ability because the majority of pre-school children, engaged in peer group play, are likely to talk in relatively short utterances whatever their levels of linguistic maturity. In addition, in play situations with peers there is little stimulation to elicit speech at the individual's limit of linguistic competence. On the other hand, we believe that if significant differences between the deprived group and their controls were found, they would be the more striking for their being obtained in these natural situations and not in formal testing. The second and third measures are indicators of readiness to communicate, and they reflect personality differences in social situations rather than cognitive developmental differences.

Average length of sentences has been used as a measure of linguistic maturity in many investigations in the past, especially in the 1930s and 1940s; they are summarized by Dorothea McCarthy (1954). She concludes that 'by three and a half years, complete sentences averaging about four words each are used. By six and a half years, the mean length of sentences is about five words'. Comparisons between the means of the main sample and their controls shown in Table 5.3 indicate that the controls had a greater mean length of utterance in every situation.

TABLE 5.3 Mean length of utterances (Total number of words ÷ number of utterances)

	Focus (N=26)		Control (N=13)		
	Mean	SD	Mean	SD	F
To another child in					
Workshop	3.41	2.11	3.91	3.06	0.36
Play room	3.72	1.69	5.79	1.59	13.45*
Workshop and play room	3.86	1.41	5.60	1.05	15.49*
To adult in					
Workshop	4.19	1.75	3.98	2.50	0.09
Play room	2.87	2.06	4.30	2.58	3.55
Workshop and play room	3.86	1.49	4.77	2.12	2.42
Workshop, all	4.23	1.64	5.24	2.16	2.64
Play room, all	3.74	1.57	5.72	1.67	13.22*
Workshop and play room, all	4.13	1.11	5.59	1.32	13.24*

* $p < 0.001$

The contrast is most noticeable in the play room while engaged in general activities, rather than in the workshop setting while concentrating on activities at the table. It is also more obvious in addressing other children rather than adults. McCarthy's analysis affords authority for the view that the contrast of mean sentence lengths - 4.1 words and 5.6 words for the deprived children and their controls respectively - represents a developmental interval of at least two years. The other two measures, mean number of words per unit time, and mean number of utterances per unit time, did not show significant differences between the two groups in any setting.

In general this seems to support the view that it is their ability to use more extended and complex sentences which distinguishes the control group from the deprived group, and not a greater readiness to communicate orally. However, certain features need to be discussed of the findings which are not revealed by statistical significance tests. On average, both groups spoke more frequently in the play room than in the workshop; but whereas the control group spoke more frequently to other children than to adults, in the deprived group this tendency was reversed. It may reflect the greater dependency of the deprived children and may confirm the findings in the behaviour observations that they were less integrated with their peers and their capacity for reciprocal activities was lower.

Chapter 6

DELINQUENCY

One of the most important findings presented by our research was the identification of non-delinquent families among a sample whose circumstances and geographical location were associated with high rates of delinquency. Recent studies have been of two types, one concentrating on the high-risk type of family, and the other on delinquency as an urban phenomenon. West and Farrington (1973), in a study of the first type, investigated boys attending six typical primary schools in a working-class area, and found four family attributes of particular significance: low family income, large family size, parental criminality, and poor parental behaviour. The importance of these factors has been largely confirmed independently (Wilson, 1975). To these four family attributes West added a fifth individual factor: low intelligence. Of sixty-three boys who possessed three or more of these adverse attributes, thirty-one (49 per cent) became delinquent, and twenty (32 per cent) recidivist. Nevertheless, West states 'Even so, this degree of predictability did not make it possible to forecast with certainty that any individual boy would become delinquent. The majority of future delinquents did not come from the small group identified as vulnerable.' (West and Farrington, 1973, p. 190). The second type, prevalence studies, on the other hand, makes estimates of the rate of delinquency of the relevant age group in different areas. The rates in the highest-delinquency areas of British cities are ten times those of low-delinquency areas (Power, Benn and Morris, 1972). High-delinquency areas are characterized by high unemployment rates, high population densities, overcrowded housing, unsatisfactory housing conditions, preponderance of early school leavers, and a preponderance of manual workers (Wallis and Maliphant, 1967). The families of our sample, therefore, are exposed to a doubly definable high risk of delinquency, through their family circumstances and their place of residence.

Clearly delinquency does not begin with the act or acts first recorded by the police, and equally clearly a large number of delinquents remain undetected, as is indicated by consistently low detection rates. How many escape the notice of the authorities altogether is impossible to estimate; even if self-reported delinquency were fully accepted our boys were too young (10-11) when

Chapter 6

tested to have committed many officially delinquent acts. The
mothers were specifically asked whether the boys had been in
trouble at school, with the neighbours, with the police, or at
home. Their accounts, reported in chapter eight, indicate that
about two-thirds of the boys had engaged in some kind of delinquent activity by age 10-11, whereas only ten boys (eleven if a
direction under the Education Act is included), just over one-
third, were officially delinquent three years later. This chapter
is concerned only with official delinquency.

METHOD OF INVESTIGATION

In the city in which this research was conducted it is the practice
of the police to caution children who have been caught committing
an offence, provided that they have not been previously caught
offending. This caution is recorded, and the records were made
available to us along with those of actual convictions. For the
purposes of this chapter 'delinquency' refers to any of the four
following categories, unless stated otherwise:
1 an official caution;
2 a finding of guilt in a juvenile court;
3 a care or control order;
4 a direction under the Education Act.
As in an earlier study (Wilson, 1962), four measures of delinquency
rates were ascertained:
1 the rate of convictions as a function of age;
2 the rate of offenders as a function of age;
3 the percentage of boys or girls so far unconvicted who become
  delinquent at each year of age;
4 the prevalence rate, being the percentage of boys or girls at
  each age who have become delinquent before or during the year
  in which they attain that age.
For the calculation of rates 1 and 2 it is sufficient to know the
total number of children who have passed through each age, and the
number of convictions and/or cautions at each age. For the rate of
entry into delinquency, 3, the number passing through each age has
to be reduced by removing those who have already become delinquent,
and by counting only those who have attained the stated age. The
data relating to age, sex, and delinquency are given in Table 3.1.

THE FINDINGS

The fifty-six families of the main sample contained 182 children
aged ten or over at the date when the search of records was closed.
The distribution by sex was as follows: 108 boys and 74 girls. The
excess of boys over girls is explained by the fact that twenty-eight
families were chosen to have a boy aged 6-7, and a further twenty-
eight to have a boy aged 10-11; when this is allowed for the preponderance of boys is not significant. The effective number of
children aged ten or over decreases rapidly with increasing age:
there are twenty-four children born in 1962, but only five born in
1952. Since one of the conditions for selection was a pre-school

child, this distribution of ages is to be expected; most of the children were under ten in 1971. Five families had no children of either sex aged ten or over, and the oldest of another four families had just reached the age of ten. Taking off the five families with no children at risk, we are left with fifty-one. Of these, twenty-seven had delinquent children, and twenty-four had none.

It is important to take into account the differences in delinquency rates between the sexes. Nationally these are in the ratio of five or six males to one female. In our sample it proved to be approximately three to one, but at the peak ages of 13 and 16 years it was two boys to one girl. As the total numbers of convicted girls are small, there are considerable fluctuations, and the average ratio can be no more than an approximation. The detailed figures are given in Table 6.1. It can be stated with some certainty that the chances of a girl in our sample becoming delinquent are clearly lower than those of a boy. Of the twenty-seven families with delinquent children, twenty-four have delinquent boys and some of these have also delinquent girls, but there are three families with delinquent girls only. In a comparison of families with and without delinquent children three factors are of importance: the number of children at risk, the ages attained, and the distribution of the sexes. These factors account for part of, but not all, the different incidences of delinquency. The delinquent families had 112 children at risk, an average of over four per family, whereas the non-delinquent families had seventy-one children at risk, an average of just under three per family. Second, in the delinquent families the number of boys aged 13 or over was thirty-seven, whereas in the non-delinquent families only sixteen boys had reached that age. Third, there were sixty-eight boys in delinquent families, an average of 2.5, and only thirty-nine in non-delinquent families, an average of 1.6. In other words, in the families who (by the time the search of records was closed) had produced delinquent children, there were more children at risk, more boys at risk, and more boys over 13 years old. Thus one might conclude that the prognosis for the families so far non-delinquent is unfavourable, especially in view of the fact that the ratio of boys aged 16 or over who had become delinquent at some stage was over 60 per cent. (Table 6.1, column 6). However, as described in chapter 11, there are significant differences between child-rearing methods of the delinquent and the non-delinquent families which probably account for the differences in delinquency patterns, and thus the outlook may be cautiously hopeful. Even though delinquent families are exposed to greater risk in terms of numbers of children at risk, ages attained and distribution of the sexes, a calculation of the risk factors in terms of 'child-years at risk', expressed for boys and girls separately, showed that families who applied strict measures of chaperonage attained lower rates of delinquency than expected in terms of children at risk (chapter 11, p. 174).

Chapter 6

TABLE 6.1 Delinquency data for the focus-family children. 'Offences' include both cautions and convictions. For a more detailed analysis see Wilson (1975).

	1	2	3	4	5	6	7		8	
	Age	No. at risk	No. not yet delin-quent	No. first delinquent	Per cent first delinquent	Cumulative per cent delinquent	Offenders		Offences	
							No.	No. per 100	No.	No. per 100
Boys	10	108	108	3	3	3	3	3	3	3
	11	92	89	11	12	15	14	15	19	21
	12	75	63	5	8	22	13	18	15	20
	13	54	43	7	16	34	11	20	14	25
	14	42	30	3	10	42	7	16	9	21
	15	36	24	5	21	54	7	20	11	31
	16	30	15	3	20	63	10	33	15	50
	17	24	11	0	0	63	6	25	8	33
	18	19	6	-	-	(68)	4	(21)	5	(26)
Girls	10	74	74	1	1	1	1	1	1	1
	11	66	65	3	5	6	3	5	3	5
	12	52	48	1	2	8	2	4	2	4
	13	46	42	2	5	12	4	9	6	13
	14	39	33	2	6	18	3	8	3	8
	15	26	20	0	0	18	0	0	0	0
	16	20	15	2	13	29	3	15	4	20
	17	15	9	0	?	29	0	0	0	0
	18	9	6	-	-	(33)	0	0	0	0

80   Chapter 6

DISCUSSION OF STATISTICAL DATA (Table 6.1)

The first two columns show the number of boys and girls 'at risk' between the ages of 10 to 18. Column (3) gives the numbers at each age who have not yet become officially delinquent; for instance, of 75 boys aged twelve, 63 had not at any time had a caution (implying an admission of guilt) or a conviction. Similarly, of 52 girls aged 12, 48 had not become officially delinquent. Column (6) expresses the same information in terms of a cumulative rate, the calculation of which is explained in Wilson (1975). This shows the steep rise from 3 per cent at age ten to 68 per cent aged 18 for boys. For girls the cumulative rate is less steep, but nevertheless by the time the girls were fifteen years old, 18 per cent had become delinquent at some stage in their lives. Because of the small number of girls involved, and the lower rate of delinquency among them, these figures can only be regarded as approximations. The conceptualization of delinquency in terms of a 'labelling process' may be criticized from many points of view, but for purposes of presentation of the material it is convenient.

Columns (4) and (5) give the numbers of boys and girls first delinquent at each age and express this information in terms of a percentage of those not yet delinquent (column (3)). The peak age of entry into official delinquency appears to be for boys between the ages of 15-16 with one-fifth entering; and for girls 16 years when 13 per cent entered (but owing to the small numbers involved this can be no more than an approximation). Column (7) gives the number of boys and girls cautioned or found guilty or brought to court for a care or control order at each year of age, and it expresses this information in terms of the percentage of children at risk. This column thus shows the total numbers and percentages brought to public attention annually, and it includes some who have previously offended, and others who are first offenders. Column (8) gives information about the number of times each year that offences were committed, thus showing the fact that some children are cautioned and/or appear in court and are found guilty more than once in one year. For instance, while eleven boys aged 13 were cautioned or found guilty or had a care order, the number of offences for them was fourteen and for ten boys aged 16 the number of offences was fifteen. Offences were counted as a single charge, and further cases taken into consideration were not counted.

The information presented in Table 6.1 gives an indirect indication of the problem of the recidivist. Counting only the major charge, and not further cases taken into consideration, thirty-eight boys were responsible for 107 offences, and twelve girls were responsible for nineteen offences, representing a considerable amount of trouble to the community. The group giving most concern are nineteen boys and three girls who were responsible for three or more offences each. These twenty-two children belong to twelve families which contain also another nine children who had committed one or two offences each. Thus twelve families (under one-quarter of those with children at risk) have produced 62 per cent of the delinquent children in the entire sample. The child-rearing methods in delinquent and comparison with non-delinquent families will be discussed in chapter eleven.

TABLE 6.2 Offences and care or control orders classified by sex

Type of Offence	Boys Convicted	Boys Cautioned	Girls Convicted	Girls Cautioned
Larceny	27	17	4	2
Breaking and entering	38	3	-	-
Taking and driving away	1	-	1	-
Drunk and disorderly etc.	3	-	-	-
Direction under Education Act	13	-	9	-
Care or Control Order	5		4	

The offences, classified by type and sex, and information about Care or Control proceedings is given in Table 6.2. Breaking and entering and larceny are the commonest offences for boys, whereas girls show a preponderance for school attendance problems in adolescence which are frequently associated with other social problems.

COMPARISON OF THE MAIN-SAMPLE BOYS AND THEIR CONTROLS

As explained in chapter three, two groups of controls were selected on the basis of (a) a mild, and (b) a moderate degree of social handicap, whereas the main sample had been found to be severely socially handicapped, although not selected on the basis of this criterion. A comparison of delinquency in the three groups is likely to throw some light on its social correlates. It was decided to define delinquency for this purpose as consisting only of convictions, and not to include either cautions or directions under the Education Act. Through the method of selection each of the two controls was very close in age to the corresponding boy in the main sample. By the time that they were approximately thirteen years old, seven boys in the main sample had been convicted of at least one offence, and one control in the moderately socially handicapped group had been convicted. There were no convictions in the mildly socially handicapped group. The comparison is set out in Table 6.3.

TABLE 6.3 Offenders: main sample and controls

	Delinquent	Non-delinquent	Total
Main sample	7	22	29
All controls	1	60	61
Total	8	82	90

($p = 0.0013$ on the Fisher 'exact' test)

It is worth emphasizing that no mildly socially handicapped control had a record; among moderately socially handicapped controls one was convicted and a further three had had cautions. Among the twenty-nine main sample boys, three had had cautions in addition to the seven who had been convicted. The difference between the main sample and the controls is statistically highly significant. The social-handicap instrument thus identifies degrees of delinquency proneness among boys aged up to thirteen years resident in poorer working-class areas.

DELINQUENCY AND ADJUSTMENT IN THE CLASSROOM

In the summary of West's investigation with which this chapter begins, low intelligence is given as one of the five important factors in delinquency. West also discusses 'conduct disorders' as assessed by a teacher-completed scale combined with ratings by psychiatric social workers. Since our investigation included assessment of intelligence by a battery of psychological tests, and of conduct in the classroom by a behaviour-rating scale administered by the boys' teachers, it is of interest to ascertain if delinquency can be largely explained by, or significantly associated with, assessable individual characteristics. It must be said at once that the number of delinquents in our sample is small, and that no great statistical significance can be expected. The range of ability within our main sample is narrow, and the intelligence of most boys is below average. There is a trend in the direction to be expected from West's work, but it is far from reaching a significance level. For behaviour the situation is different. As described in chapter four, and also by Herbert (1973, 1974b), three indicators of 'poor adjustment' are available: anxious behaviour (A); withdrawn, overquiet behaviour (B(I)); and aggressive, acting-out behaviour (B(E)). A summary of the items appears in Appendix B. The number of boys obtaining scores indicating 'poor adjustment' on each of these indicators is shown in Table 6.4.

TABLE 6.4 Adjustment rating and delinquency among focus boys (cautions and convictions)

	Favourable	Adverse*	Total
Delinquent	3	7	10
Non-delinquent	14	5	19
Total	17	12	29

(Probability on the Fisher 'exact' test is 0.03.)
* The criterion for inclusion is a score of 1½ standard deviations on the adverse side of the mean of the pooled control groups.

The results show that there is a significant association between delinquency and poor adjustment in the classroom. This pattern is contributed about equally by all three of the adjustment indicators. The fact that both over-quiet, withdrawn behaviour and aggressive behaviour are associated with delinquency suggests that the correlates of delinquency, at least among the socially handicapped, may be wider than has usually been thought.

Part two

CONVERSATIONS WITH PARENTS

Chapter 7

SHARED EXPERIENCE AND THE SOCIALIZATION OF CHILDREN

Socialization is the term describing the process by which a child learns the ways of a given social group so that he can function within it. This process enables him to understand his fellows and to see the world as a meaningful reality. Sociologists differentiate between 'primary' socialization, the experiences of early childhood lived within and dominated by the values of the family, and 'secondary' socialization, the processes that introduce the child into new sectors of society carrying a higher degree of anonymity and objectivity. Primary socialization refers to the chain of experiences in which the child internalizes reality as seen by his primary group. He learns to understand what is appropriate; what is done and how it is done; he learns to absorb specific emotions in specific situations; he begins to grasp the inevitable. What follows in secondary socialization has to do with objective reality as defined by the institutions of society. The child's private world is widened to include knowledge about other private worlds and their subjectivity in contrast to and set within the public world of laws, norms, and expectations. The more artificial character of secondary socialization, dealing with inter-relationships at a level of anonymity, tends to be less deeply rooted in consciousness and is more likely to be dislodged in conflict situations. Berger and Luckmann (1967) consider socialization to have been successful when a high degree of symmetry between subjective and objective reality has been established. The greater the discrepancy between primary and secondary socialization, the more likely it is that the effect of the 'switching of worlds' is felt as a 'betrayal of the self'.

An alternative experience in meeting a world very different from that of childhood is given by Marsden (1968) when he talks about his working-class background as at once drawing and depressing him: 'My breakthrough had not been a struggle out of the working-class. It had been an endeavour to see clearly through the respectable glasses that my upbringing and education put on me' (p. 123). He was able to formulate the difference between private and public worlds, and his readiness to face the ensuing tensions consciously helped him to resolve them by deliberate identification with one reference group without turning his back on the other. But the child whose experiences at school lack a positive emotional content, or more typically include disagreeable and even hurtful experiences, will not be ready to step into that other world which he encounters in secondary socialization. If, moreover, his family has memories of equally fruitless encounters with the public world, and if neither his parents nor older siblings provide links with that world through work, he will not be motivated to give up what he knows and adapt to that which has no meaning for him. He will become the alienated personality defined by Schacht (1971) as one who feels that life is meaningless and without purpose, because he is unable to identify with what goes on around him and because he has discovered that those around him, who represent the public world, are completely indifferent to his interests. The breakthrough seems to occur only 'under the stimulus of a strong critical intelligence or imagination, qualities which can lead into an unusual self-consciousness' (Hoggart, 1957).

Parsons and Bales (1955) describe the role of parents in terms of training their children to be fitted into their social world. Through the process of 'role training' the child gradually internalizes parental values. In close parent-child interaction a relationship is established which makes the learning process an acceptable one. Corresponding to the learning process is the mechanism of rewards and punishments which parents operate to make the child conform to the reality of his situation and not to deviate too far from parental expectations. The Parsonian model of socialization is more appropriate for the structurally isolated and mobile small family, typical of the middle classes in western, industrialized societies. 'Role training' implies that parents have expectations concerning the position the child is likely to hold in adult life; thus the parental role contains an orientation of purposefulness. The child may enter a particular career, or he may take over his father's business, or more generally, he may aim at college or university. But, as Harris (1969) has pointed out, a child growing up in a social class which is representative of a much lower level of skill and which is less subject to job mobility and geographical mobility, is not likely to be subjected to role training as defined by Parsons. Furthermore, being structurally less isolated, he is more likely to experience the influences of an extended kinship system. We expected the parents in our sample to be less consciously engaged in the processes of socialization which concern themselves with preparation for occupational roles; we had no hypotheses concerning the influence of kin.

The identity of a child is defined by membership of his family. There is no problem where he belongs; he imbibes his parents' feelings and interpretations of events, he shares their anxieties and their joys. It is, therefore, important that the observer is ready to record detail of daily activities including its critical events to obtain an impression of the variety of home atmospheres presented by a sample of fifty-six families. In ascertaining socialization processes we needed to know the degree of contact with neighbours and with relatives; and their importance as succouring or dependent agents in family life. Linked with the social network was the families' experience of mobility and loss of contacts. We recorded the families' recreational activities; daily routine:

the toys and other articles used by the children in play; parental methods of discipline during interviews; their household equipment; and critical events such as illness, accidents, loss of job, the offer of a job after spells of unemployment, desertion or reunion. The observations were transformed into indices by simple scoring techniques, and these were used, when appropriate, in presenting our findings.

But in trying to understand the world of the children that were studied we needed to know a great deal which we could not expect to witness. On first making contact with the families we asked if they were willing to take part in a joint endeavour to explore what it is like to raise a family on a low income and in inadequate housing. Parents were thus prepared to talk. We established a good relationship by weekly visits of one of us (HW) to fetch the pre-school child for playgroup sessions. Conversations developed naturally through this initial contact, and they became central to our understanding of the children's world.

The conversations took place in the main room, usually the kitchen. The mother was present on all but one occasion, the father on many, the final interview was arranged to include both. Annette Holman was the principal interviewer, on occasion one of us (HW) interviewed in her place, and the final meeting included both of us. At three of the meetings the conversations focused on daily life, its problems and pleasures, parents' memories of childhood and youth, and their links with the extended family. (The schedules guiding the conversations are reproduced in Appendix D). A fourth and fifth meeting was concerned with child-rearing methods of the pre-school child, and of the school boy. The schedules for these conversations were adapted from those used by John and Elizabeth Newson in a study of families in Nottingham (Appendix D).

The material which we obtained from the conversations has been arranged and presented in an ordered form. Parental viewpoints and experiences are described as representing typical, or minority, expressions. Clusters of practices and prevailing attitudes are combined, when appropriate, to provide family profiles, and these are related to circumstances. Many parental statements are quoted verbatim to enable the reader to take part in the conversations. We hope that in interpreting parental contributions to the conversations the meanings have not become distorted. If misinterpretations have occurred it is due to the distance that had to be bridged between the world of the families and the world of the authors. It is not due to lack of sympathy.

THE SITUATION

When you moved in here, did you find it easy to settle in? 'It's the quarter that gets on your nerves - there's always fighting, there's fighting over the kids - it's getting worse, windows get broken, they jump fences and spoil the gardens.' 'I hate it here, people are swearing and the things they do - you have no idea.' 'I don't like the house, I'm getting that I don't care if I clean it now - I'd set fire to the place, I never had people around me like them before - where I was before we all helped one another.'

'I keep myself to myself up this corner, there are some you can
mingle with, but one clan - if you get in with them - all hell is
let loose.' 'People today aren't what they used to be, it used
to be all for one and one for all - in the streets and entries,
all were together. Now it's dog eats dog.' 'I don't get on round
here - I pass the time of day, otherwise you're in trouble if you
get too friendly.' 'It's the atmosphere in the street - if I'd
known you have three choices (of accommodation offered) we'd never
have taken this house.' 'The people next door are driving me mad -
the man next door throws water over my washing, the other neighbours
are always borrowing, soap, matches, cigarettes.' 'The people round
here are very miserable people, the kids couldn't get out in the
yard - they are always complaining. They are always fighting and
rowing, and there are prowlers hanging around at night.' 'I don't
like the bloody road, all the whites are leaving and the coloureds
are coming in - though I'm not against them.'

Not all the families expressed negative feelings; twenty families indicated a resigned acceptance of the neighbourhood: 'We
don't make friends round here, we don't bother a lot.' 'I was
lonely at first and frightened, we had to watch our Ps and Qs until
we saw what the neighbours were like.' 'I was lonely at first, I
wouldn't look at anyone, I didn't know what way they would take me.
Now I know them from top to bottom - but I wouldn't go in to them.'
Some families had lived in the area for some time and they commented
on the changes they observed: 'Things are happening in the house
next door - it's terrible for children to see, boys and girls in
there.' 'All our windows were broken one Sunday at four in the
morning by a drunk. The toilet is broken, somebody took the piping
out.' 'The next door house is empty and somebody took the wiring
out which affected our upstairs electricity, there's no light, no
heating.'

The absence of a hot-water system, of a bath or indoor lavatory,
the dampness and state of decay of the houses was the subject of
many adverse comments. In some cases this topic reached crisis
point: a flooded cellar, blocked drains necessitating planks across
the yard where the water stood ankle deep, wallpaper falling off
damp walls, bricks missing on internal walls, doors off hinges. Rats
invaded the houses which were surrounded by vacated houses prior to
demolition.

DAILY ROUTINE

At the beginning of each interview session mothers were asked:
What did you do yesterday? 'Up at six in the morning, made Father's
breakfast, tidied up. Got the children up and gave them their
breakfast, washed them and saw them off to school. I waked Catherine
and took her to the doctor, I had to be there for two hours before
I was seen. I had a snack and started on the rooms, made the beds
and washed the floors. Up and down with Catherine - she wets
terribly and likes a lot of drinks. At four I give the children
their tea and start getting the dinner ready. He gets in around six.'
'I got up late - not feeling too good. The children were in bed till
twelve. Cleaned and mopped and did dishes. Children got dressed,

had tea and toast. A mate came down and we had a natter. Got tea. He (i.e. the father) went to the doctor, I sat with the children, Judy was crabby so I nursed her.' 'Got up, put water on for washing, made a cup of tea, lit the fire. . . .' There are references to child-minding functions indicating a degree of irritation or resentment: 'I was in and out all the time to see where Bernard was.' 'Marie is always following me - if I say, I'm going out, "I'm coming" she says.' 'Before tidying I tell them to sit on the settee and not get off - they'll do that.' 'I had the flu and did no housework, I had a cup of coffee and the children played around and stuffed themselves with sweets.' 'Steve wets the bed, I have to change sheets every day.' Other accounts are more contented: 'I sat at the doorstep watching Johnny play.' 'I went to the park with a flask and sandwiches - they played and I sat and enjoyed the sun.' 'In the afternoon I had a game with the children, Jaqueline likes a story.' Then there are references to poverty: 'I have to cook on the fire - the gas is off since last June.' 'We went to the rag shop with George to see if we could get any money for old rags and clothes.' 'The school inspector came, he said he'd try and fix me up with some clothes.' 'I buy more or less the same things every day, we live hand to mouth during the week.' 'In rent week we live day to day.' 'I would enjoy the shopping more if I had more money.'

By teatime lots of mothers feel they've had a full day - do you find you can have a rest and maybe a talk with your family, or does it not work out that way? 'It all depends - sometimes I can - but not when the children are playing up.' 'I'm glad to get them to bed. I find about three of them want to talk to me, the older ones bring their books to read.' 'Yes I sit and talk and I do quizzes with them.' 'I don't talk, I fool around with the kids.' 'They talk a lot, I try and listen.' These are the mothers who have sufficient energy to be alert to their children's needs. They represent about a quarter of the sample. 'You can sit down at times, but usually the kids are still in the wars.' 'I'm more or less going until I go to bed . . . I sit and doze off.' 'It doesn't work out - I'm on the go getting tea and getting them ready for bed.' 'I have to get the twins to bed to get a rest - one starts, then they all start - I shut them up - I don't want to hear them.' 'I feel if only someone would come and take them away - I don't "sit and talk": you talk to one and everyone else is talking at the same time.' 'No, it doesn't work - I just say "yes, yes" or "shut up and watch the telly".' These comments indicate the pressures of family lived in cramped quarters. Although language is used, the usefulness of language is minimal.

Sometimes having the children around all day can be a bit of a strain - do you find this at weekends? Do you feel the same during school holidays?

TABLE 7.1 Mothers' feelings about daily routine and about holidays

	Coping		So/so		Not coping		Don't know		Total	
	N.	%	N.	%	N.	%	N.	%	N.	%
Daily routine	34	61	12	21	7	13	3	5	56	100
Holidays	11	20	24	43	18	32	3	5	56	100

As shown in Table 7.1, well over half the mothers answered in a cheerful manner, or in a manner indicating that they accept their obligations and can cope. During school holidays, however, only one-fifth felt they can cope. The coping mothers typically confined themselves to short statements like 'It isn't too bad', or 'They don't seem to bother me.' 'No, I don't mind, I've had eight children round me when my sister-in-law left her family.' 'I'm all right if they don't start bickering.' 'They're all right. Holidays are not much different - I do less housework, and the girls are at home and they help me.' During school holidays many expressed very different feelings indicating much stress: 'We all dread holidays in this yard, the yard's stinking dirty. I'm scared they'll fall through the bloody floor boards in empty houses - there's no play centre, no nothing, the nearest park is.... I'd rather have them at school.' 'They get on my nerves, I can't keep them quiet in the house... if the weather is nice they are always out, but when they are indoors they play games with one another, wrestling and fighting.' 'I can't keep the place clean, I can't cope when they are all around, I had to go and get a tonic.' 'I hate the weekends, they get on top of you. Summer holidays are too long, it's murder.'

HEALTH, DOCTORS AND CLINICS

There are no adequate data on the distribution of morbidity rates by social class. Our findings are based entirely on the accounts given by parents and they do not allow a summary statement of the state of health of our sample in comparison to other groups. The accounts given in answer to our question *Does anyone in the family suffer much from illness?* are insufficient for an enquiry into health conditions, yet we feel they are of value as giving a first-hand account of parental experiences, and they provided the starting point of a series of further questions concerning the use of medical services, and the support given by the social services in times of crises.

In forty families all, or most, members of the family were reported as having had much illness, or as suffering from defects or conditions which affected their activities. Respiratory diseases were most frequently mentioned, followed by gastric complaints and skin conditions. Chronic conditions among adults included three cases of asthma, four of epilepsy, four of pulmonary tuberculosis. Five women suffer from migraine; and six were being treated for anaemia. Four fathers and fifteen mothers mentioned conditions which they variously described as 'bad nerves' or as depression. In addition, one father and five mothers told us that they had received hospital treatment for 'nervous breakdowns'. One mother had several periods in hospital for treatment of a psychosis, one coinciding with our fieldwork. Two mothers had twice attempted suicide, both by taking overdoses of sleeping tablets.

It is frequently thought that parents known to a social agency, and especially those receiving supportive care, may be of low intelligence. We considered a rating unreliable. However, three mothers told us that they had attended schools for educationally subnormal children, and at least two of the fathers did. One of the latter was described by his wife as 'brain damaged'.

The fifty-six families include almost 400 children (some of the older children no longer resided at home). In addition to the common childhood illnesses the following conditions were reported frequently: gastric infections, bronchitis, pneumonia, abscesses, impetigo, scabies, ringworm. Five children were said to have had meningitis, four tuberculosis, two poliomyelitis, and one rheumatic fever. Nine children suffer from asthma, two from epilepsy, and two are diabetic. Eight children have congenital abnormalities. One child is a spastic. Three others were described as suffering from recurrent 'fits' without detail of medical diagnosis. Five children have other apparently serious medical conditions including coeliac disease.

Among the forty families with much illness, the incidence of illness and of physical and sensory impairment is probably much higher than in the population at large, but it was not within our competence to investigate this. Obstetric problems were also frequently mentioned both by the forty families with much illness and the sixteen relatively fitter ones. The latter, although healthier on their own accounts, also had medical problems: one man was totally disabled through an accident at work and suffered from arthritis, two had asthma, one had had a kidney operation. Only four among the sixteen fitter families can be truly said to be 'healthy'.

The account of illness in the family was followed by: *So on the whole illness in your family happens a lot, or does not happen very often?* The sixteen fitter families all rated themselves as belonging to the second category. 'We are rough and ready but not ill.' 'None of us is ever took ill.' A more unexpected reaction came from the forty families with much illness; only fourteen thought illness 'happened a lot', twenty-one thought it did not happen very often, and three said illness seemed to come in clusters: 'When it comes, it really comes, but not very often.' Two families did not feel able to rate themselves.

When someone is ill in the family it is not always easy to decide when to call in the Doctor or when to go and see him. How do you do this? Do you find it easy to get the doctor? The answers are grouped into three categories, those indicating a good relationship (twenty-six families), those indicating indifference (fifteen families), and those indicating a bad relationship (fifteen families). The mothers who had good relationships spoke very warmly about their doctors: 'He's a great doctor, one of the best, he would do anything for you.' 'He's a good doctor, you can sit down and tell him your troubles.' Many of these mothers mentioned the doctor's willingness to make a home visit: 'I get something from the chemist first. If they don't seem to get better then I call the doctor. The doctor always comes - I don't phone unless they are really seriously ill.' 'He's a coloured doctor and he's marvellous, he never moans about coming out.' Three of these mothers had earlier had bad experiences when registered with another doctor: 'The doctor didn't come when the baby had gastro-enteritis. We called her but she didn't come until the morning when it was too late, the baby had died.' 'The only time I call the doctor is if they seem feverish or if it seems serious.... Dr D. was hopeless, if a child was dying he would say "Wrap him in a blanket". When Maureen had pneumonia he told me to take her down and then said she had a cold.'

The fifteen mothers who felt indifferent towards their doctors tend to turn to the chemist in the first place for advice and remedies; and some, who live near one of the hospitals, take the child straight to hospital. 'The doctor doesn't like coming out, before you sign on he tells you he doesn't want to visit. If anything happened I'd take them up the ... hospital rather than go to the doctor.' 'I mostly look after them myself, since I was ill myself I don't trust doctors, they didn't do me much good.' 'Our doctor is not one for coming out - I can't talk to him, he's not a person you can sit and talk to, he's Greek.' 'I've always got aspirins and Indian brandy, doctors can't come like they used to.' 'I have no confidence in Dr C. I've had fits of depression, I went to see her and got a silly answer, "Buy yourself a new dress and get your hair done".'

The fifteen mothers who have a bad relationship with the doctor gave accounts of the difficulties they had in cases of severe illness: 'I had to go down and plead for the Indian lady doctor to come, it was pouring with rain and I wasn't taking Denise up there with measles and bronchitis.' 'It is difficult to get the doctor to come - he says "Come down for a prescription". When Kevin had bronchial pneumonia we had to go to hospital.' 'John was coughing up blood one night, it frightened us because of his asthma. He was four years old. The doctor wouldn't come. We phoned up and Father said if he was not in the house in ten minutes, then he'd go to the police. The doctor came but he said we were to change doctors. John had pneumonia - we nearly lost him.' 'It's very rare that I go - I go to the chemist and he makes stuff up unless it's something I don't know about. Umpteen times he won't come - I had a threatened miscarriage, he wouldn't come.' This mother then went to hospital, but lost the baby. Some mothers complained about lack of real interest of the doctor: 'He doesn't examine you'; 'You get the same medicine no matter what's wrong with you.' One of these mothers tried to change her doctor but was unable to get on another list.

As shown in Table 7.2, there may be an association between the frequency rating of illness and relationship with the doctor. Among forty families who had reported much illness but believed that illness 'did not happen very often' almost two-thirds had a good relationship. However, since the data are subjective we cannot be certain about this.

TABLE 7.2 Ratings of frequency of illness and relationship with the doctor

	40 families with much illness			16 fitter families	Total	
	'Not very often'	'It happens a lot'	Others		N	%
Good	13	5	2	6	26	46
Indifferent	4	5	2	4	15	27
Bad	4	4	1	6	15	27
Total	21 (37%)	14 (25%)	5 (9%)	16 (29%)	56	100

With the younger children have you had much to do with the health visitor? Are the times the clinic is open easy for you to get along? (Prompt for whether the clinic is/was visited)

Thirteen mothers gave an account of good contact with either the child health clinic or the health visitor; six said that they visited a clinic or had done so when the youngest child was of the appropriate age, while seven reported regular visits by health visitors. The latter have hesitations about visits to the clinic: 'The health visitor wants me to take the baby to the welfare. I never took her after Jane got measles - the welfare was the only place I'd taken her, and the next day she was in a rash.' 'I never go - I only took John and he caught an infection there.' 'At 6 months Gwen had swollen glands after an injection (at the clinic) - I don't believe in them since, maybe it's wrong, but I don't.' 'I can't get there with all the kids. The health visitor comes once a week or once a fortnight and looks at the baby and has the needles for the kids.'

Twenty-six families have, or had, some contact, but there was considerable criticism of the clinics: 'They worry you too much - the baby is not heavy enough, not doing this or that - it makes you think your baby is backward, I got upset.' 'It's not easy to get down there, you have to wait in a queue, and you have to meet the other children from school - there isn't always time.' 'We don't go to the welfare - they tell you one thing, and you believe in another.' 'I went up one day with Margaret who had sticky eyes, I sat for two hours and then they told me to go to my own doctor.' In ten of these families mothers do not attend a clinic, but there is some contact with a health visitor: 'If she's in the street she might pop in.' 'She comes occasionally - she's nice.' 'She was helpful.' But there were also critical comments: 'They come sometimes from the welfare but I don't like them, some of them are not married and they demand things.' 'The last time she came she kept asking too many questions, she's too nosy - they all tell her to piss off. The coloured nurse who used to come was nice.'

Seventeen mothers were very critical of clinics and had no contact with a health visitor: 'I never see them, I had a few words with one, I was sick myself - I was sick of running up and downstairs fetching the twins - she's not much help. I asked her when I could get my teeth done, she said "I don't know" - so I don't bother her.' 'The welfare haven't been here since we came, I don't like them coming telling me what to do.' 'I never take them to the clinic, I don't believe in injections.' 'No, I don't believe in them, you sit there for two hours and then you go out and they get colds - I get the injections done by the doctor.'

TABLE 7.3 Contact with health visitors/clinic and relationship with the doctor

Relationship with doctor	Contact with health visitor/clinic			Total	
	Much	Some	None	N	%
Good	9	11	6	26	46
Indifferent	2	7	6	15	27
Bad	2	8	5	15	27
Total	13 (23%)	26 (46%)	17 (30%)	56	100

Table 7.3 relates the information we had about relationships with the general practitioner and contact with the child health services. Nine families are well served by both; the number of families, who have little or no contact with the child health services and whose relationship with the general practitioner is not good, is twenty-six.

Whatever the reasons for failure of contact may be, it is alarming that 46 per cent of the sample had little or no effective contact with the medical services. All families had at least one child under school age, but there were thirteen families who had eight or more children of whom three or more were under fives. These families have priority needs for advice and help in case of illness. However, ten of the thirteen have little, or no contact with the child health services.

Some children seem always to be getting colds, others have accidents. Have you found this with any of yours? Rather more than half the families, thirty-four in number, had experienced severe accidents; five boys and one girl were notably prone to accidents: 'Every holiday he's in the Accident Hospital, he goes on his own and then lets you know. He's cut his head open, cut his arms and foot, and ripped his bottom on the stopcock in the school toilets.' 'He's the one for accidents: he fell down a cellar, off a roof, fell on his head at school and off a tree.' 'Peter cut his wrist through running wild - twice he had broken arms, he's always in hospital, he was in on fireworks night.' Sixteen children suffered serious burns or scalding in accidents in their homes, mostly leading to operations for skin grafting; and one child lost an eye. Six children were involved in motor accidents, and nine in other serious accidents which required medical attention and, in some cases, hospitalization for a long time.

If father or mother have a serious illness or an accident it can cause quite an upheaval. Have you ever had to manage in these circumstances? Thirty-one mothers said they had not found such a situation critical; they had managed. 'The older ones look after the younger ones' was a typical comment. Some mothers carry on however unwell they may feel: 'I have to do it all myself - ill or well.' This mother did not like her husband to take time off work because they could not afford the loss of earnings. In other families accounts were given of the father's willingness to take over: 'My husband took time off when I was in hospital, we've never asked anyone to come in.' That such an arrangement can have financial repercussions is clear from the comments of three mothers: 'My husband looks after them, he won't let anyone else have a chance. He won't go to the Social Security, even when I had one of the children and he was off work looking after them.' 'Father has to manage - he's lost a couple of jobs because of this.' 'Father managed with Rita, he lost his job because of having to take time off.' In nine families the younger children had been taken into care during the mothers' illnesses. Three families rely on the help of neighbours or friends, and four families made arrangements with a relative to mind the younger children for part of the day.

Did you find that when you were having your children you could make arrangements for the older ones easily? The usual arrangement is for the father to take his annual holiday; forty-one families described this: 'I wouldn't leave them with anyone else but my husband.' 'We managed, Father looked after them, Father has been with the firm for twenty years. I clean the house from top to bottom and wash and iron before the baby comes.' 'When Wendy was born he was allowed to work the six to two shift for about eight to ten weeks. I kept the oldest girl at home as well - she's a little mother and can do anything.' Typical remarks stressing family independence are: 'We've never asked anyone to come in', 'I wouldn't like to leave the children with strangers', 'Father copes - he wouldn't let anyone in'. Several women mentioned older children who would give the father a hand, or who would run the family without the father's assistance while at work. Two mothers reported the loss of employment because the men had taken odd days off; both were unskilled labourers.

Just over one-third of the families mentioned the help of relatives; fourteen families rely on a grandmother, and another five on other female relatives. Neighbours were mentioned by only three families. In all these cases of assistance by relatives or neighbours it is the custom to take the younger children into their homes, if they live nearby, during their fathers' working hours. Seven mothers had used the social services for short-term care.

We obtained information from the social services department about the number of families and the number of occasions when children were in care. As shown in Table 7.4, eighteen families, just under one-third of the sample, had used the service when unable to look after their children. The number of occasions varies considerably, but is much higher among families who are on preventive supervision. It appears that the service is used as a last resort even by families who have many children and whose kinship network is broken through frequent moves.

TABLE 7.4 Number of occasions of children in voluntary care

	Under 1 month	1-6 months	Over 6 months
1 Preventive supervision			
Family A	3	2	-
B	6	-	-
C	4	-	-
D	-	1	-
E	3	-	-
F	4	1	-
G	19	3	-
H	2	-	-
I	42	9	2
K*	Record not available		

(continued on p. 98.)

TABLE 7.4 (continued)

	Under 1 month	1-6 months	Over 6 months
2 Short contact			
Family L	2	–	–
M	6	–	–
N	1	–	–
O	5	–	4
P*			
Q*	Records not available		
R*			
S*			

*One family on preventive supervision, and four on short term contact reported use of the service; records were not available.

STRESS

We must pause here and take stock of the stress situations about which parents have talked with us. Stress in the milieu of poverty is all-pervading in the sense that there is a chronic condition of being in want of something that is needed or desired; quite often this is an essential article, something that people are expected to have and the absence of which marks people as deficient. Sometimes it is a craving for a food, or a drink, or a smoke that can assume an overpowering force. Sometimes it is a longing to do something that cannot be afforded; to have a holiday, to have a rest, to have an evening out. All these sensations of want are felt acutely by the people we talked with, and often this feeling is sharpened when the want is expressed by one of the children whose parents are unable to meet it. Prolonged poverty means more than any of these conditions, because it brings about an accumulation of unpaid bills, for electricity and gas already consumed (and this leads to having the supply cut off), for rents due long ago, for the purchase of clothing, a piece of furniture or other major expense that had to be undertaken by borrowing. Mostly this is done by means of 'clubs' that advance loans at a high interest rate; but sometimes when a purchaser was found credit-worthy he had an hire-purchase agreement. All the families had had experience of these situations. They were shared experiences. It is the kind of situation described by Hoggart (1957) as 'a life of tightness and contriving' that in his childhood had brought about a feeling of sick envy of those who could pay their bills cheerfully, a life lived week by week in the knowledge that there is little one can do about it except keep one's dignity.

Beyond this shared experience of want there was a condition of extreme want suffered by some, but not all, the families. Twenty families were found to live on incomes that were chronically below the poverty line, some of them substantially so. This contrasts with twenty-nine families whose incomes, at the time when they were examined, were on average at the poverty line, and seven families who on average stood at 25 per cent above the poverty line. All but one of the fathers in the last group had steady employment, and had not been overtaken by illness. The group of twenty families whose incomes were consistently beneath the poverty line were thus subjected to an additional burden of stress. (For details see chapter two.)

A second additional stress, not shared by all families, consisted in the combination of large family size and an exceptionally large number of pre-school children, as discussed on page 21, chapter 2, and again on page 96 of this chapter. Thirteen families were identified as having an exceptional burden in terms of parenting, in contrast to thirteen families with up to five children of whom no more than two were under five years old, and another thirty families of intermediate size and composition.

A third, additional, stress consisted of the presence of a severely-handicapped child in the family, demanding much attention from an already overburdened mother. There were four families with such an invalid child.

A fourth stress consisted of the presence of an invalid father or mother. Parents, in their accounts of illnesses, had reported a variety of conditions that implied a degree of restriction of activities. Some of the men were registered as disabled, but none of the women were eligible for registration, since they did not seek employment. The situation is confused by the fact that disability does not necessarily mean unemployability; in fact one father, said to have suffered brain damage and a former pupil of a school for educationally subnormal children, held a steady job as a street cleaner. Nevertheless, he provided little support for his wife in bringing up the children. Another man, an epileptic who had spent his school years and late adolescence in a colony for epileptics, managed to hold down a job as a machine operator. Only two fathers provided clear cases of unemployability through an accident; the pelvis of one man was crushed under a concrete mixer; another suffered extensive injuries in a motor accident and eventually died. As suggested by Culyer, Lavers and Williams (1972), an index of ill-health would need, as a first step, a simple description of painfulness and of the extent to which activity was restricted. In the absence of medical opinion the statements of the parents were accepted; in many cases there was independent corroboration. Minor ailments of a chronic nature, not debarring the person from undertaking normal activities without serious discomfort, were not counted. Besides physical ill health, various mental states were described to us. One mother suffered from a recognized psychosis for which she was receiving treatment, sometimes as an in-patient, sometimes as an out-patient. A further twenty-five women gave accounts of mental trouble, such as blackouts, frequent giddiness, occasional fainting, depression, or nervous breakdown, for which five had received treatment. Two of these women had twice attempted suicide. Four of the men gave accounts of mental trouble, leading in two cases to very restricted activity. On the basis of parental information four men were severely physically handicapped, and another two sufficiently so to restrict their activity. Although many women complained of ailments of a chronic nature, only one, an epileptic, was severely restricted in her activities. In all, thirteen families presented an additional stress factor of physical and/or mental disability of father and/or mother.

These stress situations: poverty, family size and number of pre-school children, an invalid child, physical or mental handicap of a parent, were each counted as one additional stress in a stress score. The total stress burden for each family made the division of the entire sample into three groups possible. A 'privileged' group of twenty families was not burdened with any additional stress. Twenty-four families had one additional stress, and twelve families were burdened with two or more additional stresses.

Stress has a bearing on parental protective capacity. Depression, apathy, and what may appear to be an insensitivity or indifference to the needs of the children must be seen in relation to stress. Chronic poverty implies malnutrition, disturbed nights, anxiety. Additional stress, such as the presence of a handicapped child, a large number of very young children, a disabled or handicapped marriage partner, means an added taxing of personality resources. Was there a relationship between the number of stress situations and the atmosphere of the home? Our problem consisted in the measurement of the latter. We decided that the accounts the mothers had given of their feelings at the end of the day and during holidays, reported earlier in this chapter, would give a more accurate indication than any ratings undertaken by us. But the atmosphere of the home is not entirely determined by the way the mothers tend to feel at the end of the day, or by their feelings during school holidays; if one attempts to grade home atmosphere in terms of a degree of cheerfulness which may ease the daily battle then one must include in this a measure of parental marital relationships.

The quality of marital relations was assessed independently by Annette Holman (who had conducted most of the parental interviews) and by one of us (HW) who had visited weekly for play groups, and also taken part in interviewing. Three categories were used; two of these appeared obvious, they concern the two extreme states, and we referred to them as a 'normal' marriage and as a marriage 'showing friction'. We used an additional 'intermediate' state in cases of doubt. The index used for the marriage rating is the sum of two independent ratings. There was concordance between the two raters in forty-three cases (77 per cent), some disagreement due to difficulty in assessment in eleven cases (20 per cent), and a rating difference in two cases. When agreement was not reached, the family was given an 'intermediate' rating. We rated twenty-six parents as having 'normal' marital relations, and seventeen as showing much friction. Thirteen were given an intermediate rating. The rating of marital relations was combined with the rating of maternal feelings about their daily routine, and during holidays, by a simple scoring system to give a measure of the home atmosphere. On the basis of this we found nineteen families to have mothers who feel they can cope in daily routine and during holidays and whose marital relations were rated as 'normal'. Twelve families had mothers who said they could not cope at all and their marital relations showed friction. Twenty-five families were placed in the intermediate category. The three categories corresponded closely to the impressions which the two raters had formed on closer acquaintance with the families; we called the three groups 'happy', 'intermediate', and 'unhappy' families.

We wanted to know whether families burdened with additional stresses were measurably less happy than the rest. A correlation of home atmosphere and additional stresses, shown in Table 7.5, did indeed demonstrate what one expected. 'Happy' homes are significantly associated with families whose total stress burden is relatively lighter, and conversely, none of the families with two or more additional stresses had a 'happy' home atmosphere.

TABLE 7.5 Home atmosphere and additional stresses

Home	Additional stresses			Total
	None	One	Two or more	
'happy'	9	10	0	19
'intermediate'	8	11	6	25
'unhappy'	3	3	6	12
Total	20	24	12	56

(Chi squared = 10.9 with 4 df, $p < 0.05$.)

The consequences for children who grow up in unhappy homes containing one or more specific stress situations cannot be described by us in any measurable way, as will be discussed in later chapters. It is not the home atmosphere nor the degree of stress which affects those behaviour patterns we have been able to assess; these are governed by parental measures which are independent of home atmosphere. Nevertheless, the atmosphere of the home is bound to affect the child's personality in a profound way by arousing feelings of fear, doubt, distrust and perhaps despair, or alternatively, by giving him a degree of confidence and assurance that things will not get worse, bad though they be. In almost two-thirds of the families, accounts were given of particular situations in which mothers felt deep despair. These were recorded in answer to our question *Has anything ever happened so that you have really felt at the end of your tether? If so do you mind telling me about it?* 'I often feel like that. I have to see the doctor about my nerves, I'm depressed, very cranky and cross. It's the kids, the house, no money and the bills - it's all worry and no rest. I could do with a nice holiday, I seem to be going down all the time - I don't get no help, no-one I can talk to, I was going to tell the doctor about all the worry.... Rich people can go where they like and do what they like, but I couldn't even go home for my father's funeral.' 'Everything is piling on top of us and I can't get my head above water, it's more or less an existence from day to day.' 'I feel like that every day lately, the toilet has been out of action for six weeks, the kitchen tap out of order longer, I had to get water from neighbours - now water drips all the time on the floor.' 'Once or twice when he's brought no money and when Gwen and Joe were in hospital, he's done six weeks' work in three years, if he gets a job he only sticks it for a couple of days or weeks.... I usually throw something at him and tell him to get out.' 'Yes many a time. He is continuously going out. When he wasn't at work he would be out from ten till six, and I never knew where he was. There was no-one to talk to, I have spasms of crying all the time.' 'He says awful things like "I wish you would die", and "you'll go to the madhouse" - he aggravates me, last time I nearly hit him. He mocks me because I go to the doctor for my health, I've always been frightened of him.' 'Some days I do when the children are at home all day, when I get depressed I feel like that, I wish I could run away for a few hours.' 'Trying to manage has made me a nervous wreck. On Saturday I sent up to the butcher to let me have some meat till Monday. I'm always borrowing over the weekend - I go to second-hand shops all the while - you get fed up with it. It's always someone else's stuff, not your own.' 'Christmas is a terrible time when you've no money, and I and the children weren't well. I had the electricity cut off for about a week, then I went to the Children's Department, they phoned his boss and asked him if he would help, the boss lent him the money.' 'When my nerves were bad I took two overdoses, both times the electric had gone off, I thought there's no more to it. I couldn't keep still, the children got on top of me, father never bothered. I was trying to cook on the fire, he never said we'll try and get the electric on - he didn't share the worry. I've shown him the bills, but no effect, even on holidays he won't give you extra, he can't see they want clothes or shoes, he won't help with that, I have to cut down on food. The probation officer came down to see me that night and told me not to do anything silly - I took sixty tablets.'

Thirty-four mothers gave accounts of situations in which they felt that they could not cope; they include all but one of those rated as having an 'unhappy' home atmosphere. Some of the rest did not want to talk about particular crises, others passed over the theme lightly: 'It's money - I used to worry about money, but I've never gone to the gas oven yet.' 'Only when the kids keep running away - we have to go out and look for them late at night.' 'There are days when the children play up - I could scream.' 'I wouldn't mind if the house blew up.' But some parents accepted the situation: 'We are fairly lucky, at least as far as health goes.' 'I'm not bothered with nerves, as long as I have grub and a fire. Only when the children had nothing to eat I felt like committing suicide - I starve myself for them, I don't eat what I should eat.' For some of these women the current situation was one of release after homelessness or after a bad marriage and a new start with another partner. It was an improved situation in comparison with previous experiences.

The support given to the families in times of a crisis such as maternal illness or the birth of a baby has already been referred to on pages 96-97. When mothers had talked about 'feeling at the end of their tether' they were asked: *Most of us when this sort of thing happens feel the need to go to someone else for help and support. Who would you go to?* The families who live close enough to their relatives and who are supported by them are in the minority, and even fewer feel a desire to get comfort or advice from a relative. 'I'd go to my mother but I could not tell her everything.' 'I went to my sister-in-law to get some advice on getting a separation but she wasn't much help.' Only seven mothers said that they would go to a close relative, usually the mother or sister. Eleven mothers mentioned a friend; most women merely gave the name, others enlarged. 'A friend I've always gone to since I was seven years old; she helped me a lot in my first marriage.' Four mothers mentioned their doctors, and ten a social worker. Twenty-four mothers did not go to anyone outside the family, either because there was nobody about: 'the friend I used to go to has been rehoused, I just sit there with a long face till it wears off'; or because independence was preferred: 'I suppose my neighbour would listen - she's not really interested - but I like to be independent. We (i.e., the father and the mother) talk it over together, he says "something will turn up" - I always look on the black side.' 'I wouldn't talk to anyone, they would gossip.' 'No, I don't bother, I sit and have a good cry and that's that, I get depressed, but I get over it.' 'I've always worked it out for myself.' 'There's no-one, I usually sit and cry.' And one mother said 'I'd rather write it down than go to someone.' She had in fact written some notes about her Irish childhood and about a critical period in her marriage when she planned to leave the family.

We recorded two kinds of contact: first, practical help such as child-minding in times of crisis, or giving a hand during illness, or a loan of money or food or other articles; and second, support of a more general nature, such as is derived from friendship or a trusting relationship with a relative. The records are based on parental accounts and on observations made during weekly visits. The degree of contact was rated in terms of none, occasional, and frequent, and the maximum score possible was eighteen. Just over half the families, twenty-nine in all, scored eight points or more; these are described as being members of a relatively close-knit social group. Just over a quarter of the families, eighteen in number, had a relatively loose-knit network, scoring five to seven; and nine families (six of them Irish) were very isolated, scoring one to four. A correlation of degree of happiness of the home atmosphere and the closeness of the social network showed a chance relationship, as might be expected. Motivation for seeking closer contact, and in fact establishing social contact, varied from case to case; some people preferred independence, some longed for closer contacts but distance from loved relatives prevented this; some families sought and found support among their kin or among friends. Only one of the twelve families rated as having an unhappy home atmosphere was also very isolated; five of them had a close-knit social network.

Thus the children, in the process of growing up, have many shared experiences. They live in overcrowded conditions, being members of large families; their homes are inadequate by current standards; the neighbourhoods are rough and disliked by most who have to live in them. They experience poverty, by which we mean that they go short of things considered essential or normal by others around them, and they lack experience of situations encountered as normal by others around them. Most, if not all, the children have first-hand knowledge of illness, disability, accidents and mental stress expressed in a variety of symptoms. They must learn, in growing, to come to terms with or to contain the situations of stress to which they are subjected day by day. As Bruner (1966) puts it: 'Growth depends upon internalizing events into a storage system that corresponds to the environment. It is this system that makes possible the child's increasing ability to go beyond the information encountered on a single occasion.' Events in these children's lives repeat themselves and form a pattern and thus become part of what appears to them to be the hazardous business of living. Events are internalized, and with them also the feelings of other members of the family, especially of their mothers and their fathers. Their parents' chance remarks, moods and outlook on life are part of the running commentary by which the children learn to accommodate themselves to what life brings them. The children's capacity to internalize events enables them to accept as normal a situation that, by any standard other than their own, would be considered extraordinary or tragic. It has been suggested (Lewis, 1967) that the 'culture of poverty', however much it may deviate from the norm, has variety and richness within its own limitations. We are not convinced that the lives of poor families in Britain constitute a 'culture' or a 'subculture', but we would argue that the variety of life-styles encountered among these families contains a tragic aspect. The draining of human energy and potential is the element that outweighs all others and has an overpowering effect on the growth of children that no other element can counterbalance.

Chapter 8

THE WORLD OF THE CHILD

One of our primary objectives was to study the methods of child-rearing used by the families who formed the main sample. Child-rearing involves two distinct objectives, one being the acquisition of 'techniques', the other the training of social behaviour. The former 'equips the child with what may be called "problem-solutions" - it teaches him how to carry out his intentions, how to reach the objectives which are desirable to him' (Trasler, 1968, p.1); it also involves the acquisition of language, the recognition and categorization of objects, and the use of tools. The training of social behaviour includes the complex process 'by which the child develops those emotional reactions of admiration, aversion, guiltiness, and so on, which will underlie and motivate his social behaviour' (ibid.). It is this latter aspect of child-rearing processes in which we were primarily interested.

The classic study of parental practices of child-rearing in the USA by Sears, Maccoby and Levin (1957) describes how a sample of American parents carried out their activities. The study included 379 mothers of 5 year old children. In a set interview session mothers reported their feelings about motherhood and their families, they described their child-rearing practices from the child's birth until he was 5 years old, and they gave some account of the child's behaviour which enabled the authors to determine the effectiveness of some of their practices. The aspects which were considered most influential in the socialization process were the way in which the mother disciplines the child in her effort to get rid of undesirable behaviour, the degree of tolerance and consistency of control that she applies, her temperamental qualities, and the values she holds up to her child in the inculcation of more mature behaviour. The participation of the father in the child's activities, and parental concordance of objectives in child-rearing were also ascertained. The findings were related to social class, and significant differences in some practices were recorded.

A later study of parental practices in Nottingham (Newson and Newson, 1963, 1968), although similar in many respects, improves on the American study in at least two ways. The study focuses on the child at a specified age, thus avoiding parental recall of events. In addition, an attempt is made to examine 'the total climate of the child's upbringing' (1968, p. 18) rather than concentrate on single measures and their supposed effects. For this purpose a series of indices were developed by which attitudes in general could be measured. For instance, in the study of the 4 year old (1968) two indices are introduced, one to measure 'child-centredness' and the other to measure 'evasion or distortion of truth'.

In the two studies which describe infant care and care of the 4 year old (Newson and Newson, 1963, 1968) the authors demonstrate considerable variety in parental understanding of what is 'right' and 'wrong', desirable and undesirable, good and bad in the upbringing of children. In addition, parents admit that they do not in fact always practise what they would like to practise even though they may have strong and clear ideas about desirable methods of raising children. Sometimes these variations coincide with class differences. The sample had been selected to be representative of the Registrar General's social classes, with a division of class III into manual and non-manual. The majority of middle-class families (64.5 per cent) were found to have two children or less; whereas the majority of class V families (57 per cent) had four children or more. The authors point out: 'This factor of family size needs to be borne in mind, together with other background differences when class comparisons of behaviour are made' (1968, pp. 41-2). The earlier American study had also taken account of the importance of family size:

> Although the working-class mothers were somewhat more severe with their children than were middle-class mothers, this difference may have had nothing to do with social-class level per se. Perhaps the problem was that the working-class mothers had larger families and less help in caring for them so that they did not have the time to give much individual attention to their children, and had to be stricter in order to cope with the many demands the children made (Sears et al., 1957, p. 435).

In their study, however, no appreciable difference was found between young working-class and middle-class families in the number of children they had, which was 2.5 and 2.7 respectively. Nevertheless, the point remains a valid one. First, working-class mothers are likely to have less help in caring for their children, and second, young parents tend to handle their children in the way they were handled themselves, and there is a carry-over of traditional practices from previous generations whose experiences were rooted in the large family.

The American study does not refer to the quality of housing as an important factor affecting child-rearing methods. The Nottingham study, on the other hand, points out the relevance of housing; in a discussion on parental control of aggression the reader is reminded that

> in a terrace house the woman standing at her own back door is also standing at her neighbour's. If trouble is brewing, there is not much chance for it to simmer down when the opponents are continually brought face to face with one another, and when they can never even escape from each other's voices.... The council estates, with their separate gardens and, in particular, their separate garden gates, bring to working-class wives the possibility of selecting their children's play-company (ibid., pp. 130-1).

There are several references to the convenience of having a lavatory in the house when it comes to toilet training, and of having a bathroom:
> Difficulties are especially likely to occur, of course, in houses which have no bathroom: the mother may well choose the daytime, when the older members of the family are out, as the only time in which she can secure enough privacy to wash all over; and if she feels that modesty is necessary before her pre-school children, this will certainly curtail her freedom of activity (ibid., p. 369).

One of the most interesting findings is the following:
> We took the working-class sample only and compared mothers living on estates with those in the more over-crowded central areas. A significant difference is found in that *frequent* punishment is more likely in the more overcrowded districts (ibid., p. 424).

There is also a reference to the role that improvements in housing conditions have played in aiding families 'to find fun in parenthood' (ibid., p. 521). Thus both factors, size of family and quality of housing, were found to play an important part in shaping child-rearing methods. Both are selection criteria in our study; they contribute to differences we found in the way parents handled their children in comparison with parents in the Nottingham study.

Many of the child-rearing practices in the Newson study were found to be related to social class. Traditional working-class practices are described as tending to be rather authoritarian and middle-class practices as favouring a more egalitarian regime. Repressive measures more typical of the working classes are contrasted with a rational flexibility of the typical professional mother. But there is a wide range of attitudes in all classes. Where there are class trends they are sometimes diminished by reference to generational changes. In their first study (1963) the authors report changes in terms of a 'climbing down from the parental pedestal of a generation ago', and in their second study (1968) they state: 'the change in parental attitude continued to be emphasized'. Parents hoped 'to achieve a closer relationship with their own children' than they had had with their parents (ibid., p. 522). Thus there emerges a time dimension which becomes a qualitative dimension; certain practices are felt to be out of step with modern insights. The authoritarian-democratic dimension and concomitant practices in the use of language is added to by the authors in considerably developing a concept much discussed in the literature dealing with child-development, which is 'child-centredness'.

The Nottingham study is designed to interview mothers at regular stages of child development. By kind permission of the authors we were able to use the schedules for the 4 year old, the 7 year old and the 11 year old. In view of environmental conditions and family circumstances of our sample certain modifications were indicated. Interviews of up to 200 questions would test the staying power of most of the mothers too severely. The interviews were to be enjoyed, not endured. They were to be administered in the families' living rooms and, very probably, in company of several children who might present a distraction from their mothers' endeavours to concentrate on topics they do not normally discuss. The most obvious category of questions to be omitted without loss to our general understanding

of their viewpoints was that which deals with generalizations or
principles of child-rearing. These, we hoped, would become clear
in the ascertainment of mothers' descriptions of specific situations.
We believe that, in fact, we have succeeded in doing this. Omitting
questions on principles had the added advantage that mothers who
rarely or never thought about child-rearing methods in the abstract
were not embarrassed by questions of this nature. A second category
of questions which was omitted ascertained activities which are untypical for families living in poverty. They referred to family
entertainments, such as visits to the theatre, concerts, exhibitions
or art galleries; and to the provision of private tuition in music,
dancing, etc. The decision to omit this category was made in the
knowledge that a small proportion of families might be missed out;
in fact rare occasions of a visit to the pantomime or a museum were
reported when family activities in general were discussed. A third
category, asking questions about behaviour at meal times, was omitted
as we knew that almost all the families had to take their meals in
relays, because they did not possess, nor was there space for, a
table and chairs to provide seating for all members of the family.
The questions which were retained in our schedules are listed in
Appendix E. We selected for presentation in the text those questions
which we believe to be most significant in describing parental child-
rearing methods in our sample in contrast to the Nottingham sample;
the findings are presented together with the Newson findings. Thus,
fifty-six 3-4 year olds and twenty-eight 6-7 year olds are compared
with the equivalent age groups in the Nottingham study. We also
used the Nottingham schedules for the 11 year olds, but as the findings concerning this age group were not available at the time our
findings were analysed, the practices of the parents of 10-11 year
old boys could not be compared with the Nottingham sample, but had
to be presented by themselves, in relation to the practices of
younger siblings, and of other families with younger school boys.
The families with 10-11 year old boys contain one who has also got
a seven year old in the sample; the total number of the older age
group is thus twenty-nine.

THE CHILD AT PLAY

Fifty-six children aged 3-4 years

The Nottingham study found that most 4 year olds are not content
to be left on their own for very long, and if restricted to their
own homes a great deal of skill and patience is required to keep
the child happy. The contrast with the families who took part in
our study is immediately obvious. Seventy per cent of the group have
more than one pre-school child, thus even during school hours the
under-fives are not left to their own company. The seventeen pre-
school children who did have this experience while their older siblings were at school did not like playing on their own. Mothers
told us: 'She wanders about when the others are at school.' 'He
likes someone to play with him, there's so many children round here,
there's always someone.' 'He's never in - he wouldn't play on his
own.' However, we discovered that three mothers restrict the child's

play with neighbours' children: 'We won't let them outside - she's had children in the yard. When she's on her own she's all right, if other children come she screams, she likes her toys to herself.' 'When the others are home he plays with them. The children round here - no thank you, they are robbers.' 'She plays outside, she plays with one of the neighbours' children - they hit her though, she's better on her own.'

Taking all pre-school children together, well over half (57 per cent) were described as not liking it at all when they have to play by themselves. This contrasts with working-class children in Nottingham, 23 per cent of whom prefer never to play on their own. Unwillingness to be left to their own resources is probably due to the absence of toys and other equipment suitable for play. The fact that only 7 per cent of children from middle-class families in Nottingham prefer not to play alone, bears out this guess.

The Newsons found that most mothers in Nottingham were able to spare money for toys of lasting quality and of good design, which enabled them to state that 'the average ten-year-old, unless he has given or thrown away his outgrown toys, has a collection of possessions dating back to when he was a toddler' (ibid.,p. 137). Coupled with this finding was another; the majority of 4 year olds seem to have a strongly developed property sense which comes to the fore when mothers attempt to clear out accumulations of broken toys, or when they try to give away toys they consider outgrown. The comments of our mothers drew attention to a very different world: 'They haven't got many toys, she plays with the bloody ashes, bugger for the dirt she is - she plays with Pauline when she's home, they play shops together.' 'She breaks her toys, she's not interested, she plays with handbags and old clothes - we haven't any money for toys. I don't like them playing with dirt and water.' 'They play shops and ring-a-roses and gambolling - I can't manage to buy toys on what we get. He's got a toy car he likes a lot. They had a game on the dustbins yesterday.' 'It's not easy to get toys, especially now Father is out of work.' 'They never had all that many toys, my husband says he won't buy toys because they break them. I don't look at it that way - I get them dinky cars some weeks. He likes looking for worms, he plays with a ball, and on his bike'(a tricycle).

On each of our home visits the number and type of toys seen were recorded. In thirteen homes we never saw any toys; we doubt that there were toys which were always out of sight or tidied away. In thirty-three homes we found some toys, mostly much used and damaged, scattered around the kitchen or the front room. Many were plastic; animals, cars, dolls, and so on. There were some soft animals, tattered picture books, and the occasional tricycle. In ten homes there was a range of rather better toys, including some table games like puzzles, snakes and ladders, or playing cards. Nowhere did we find the type or quality of accumulated toys which the Newsons describe as typical in most homes.

Mothers were asked *When he is playing at something do you ever join in?* The Newsons found that size of family played a role in the proportion of mothers who 'actively participated in play at the child's own level': while 76 per cent of mothers with one or two children did so, only 57 per cent of mothers with larger families participated by taking 'a special interest in certain kinds of play' and by 'sharing wholeheartedly in his role-playing games' (ibid., pp. 166-7). But mothers in class V (unskilled workers) were even less likely to participate than other mothers with larger families; only 44 per cent did. The Newsons suggest that 'these mothers often have a somewhat more formal and tradition-oriented conception of the maternal role' (ibid., p. 172). In our sample only thirteen mothers (24 per cent) participated wholeheartedly as defined by the Newsons, an even smaller proportion. Was this, we wondered, related to the absence of toys? We found that of the ten families whose children had a fair range of toys, nine mothers do participate in their children's play, although on a definition which is not quite as stringent as that of the Newsons. On that same definition, only twenty-one mothers in forty-six families with few toys or none played with their children, a significant difference (exact probability test $p < 0.025$). We had to widen the definition of participation, as there is little role-playing and other intensive forms of play in poor families; on this basis 53 per cent participate.

A series of questions aimed at eliciting the degree of restriction of play at home: *Is there any sort of play you don't allow? Do you let him make a lot of noise if he wants to? Do you let him jump on his bed and use furniture for his play - like making a train out of chairs? Do you let him make a mess playing with water or paint or earth or flour? Does it bother you if he gets really dirty while he is playing?* Almost three-quarters of the Nottingham sample impose moderate restrictions, but a significantly greater proportion of professional-class mothers are influenced by 'the idea that messy play is not only natural but right' (ibid., p. 155). The proportion of mothers in our sample was even greater, 29 per cent imposing almost no restrictions in comparison to 21 per cent of Nottingham class I and II mothers. 'I'm used to it now, sometimes it bothers me, I've just washed and dressed them, and they come back covered in oil.' But over half impose moderate and 16 per cent impose much, restriction on play.

Twenty-eight boys aged 6-7 years

Like their younger siblings these boys do not like being indoors: 'He gets fed up indoors, he says he's got nothing to do.' 'He can't sit still for five minutes.' Only one boy was described as an indoor child. *Which does he like best - playing with other children or playing by himself?* Eighteen boys (64 per cent) liked company, six preferred to play alone, and four boys liked either. While the pre-school children are more gregarious than the Nottingham children, it appears that three years later, when the children have had experience of schooling, significantly fewer than nonskilled Nottingham 7 year olds are gregarious (with 64 per cent in contrast to 82 per cent). The six 'loners' are often involved in the activities of siblings; whether they want to be alone or not, there is no choice. Two boys never play with other families; this is their parents' decision and will be discussed further on.

What about days when there is no school? What does he do with his time mostly? 'He runs in and out.' 'He has his tea, it's the first thing in his head, then he goes out playing - he hangs around, he doesn't move far from the house.' 'He plays out and watches the

telly, he's been digging today and shopping. He takes Dolly in
the push chair up and down.' 'He goes in the garden, he wrecks
the garden, he always throws things. He goes tatting round, he
goes with a rag man and sometimes on his own.'

What about fighting - does he do a lot? Nottingham mothers were
divided equally into three groups maintaining there was much, some
and little fighting, with no significant variation among the classes.
In contrast, more than half our mothers (54 per cent) reported much
fighting, and only a quarter little. Most fighting happens between
siblings. When it comes to fighting with other children, 61 per
cent reported little in line with 68 per cent of Nottingham mothers.
The mothers see fighting as natural among boys of that age: 'Him
and his friend David, they do a lot of fighting, but they are the
best of mates the next moment.' 'You'd think he was getting piece-
work rates for his fighting.' 'He boasts about his fights.'

Six boys were described as very peaceable, fighting little with
siblings and hardly ever with neighbours' children. 'Paul always
gets on with other children, he never gets roused.' Joseph has a
severe speech defect and is described as a placid child. Steve is
obese and does not mix. Alex is shy, 'he's all for me, he does
the dishes and makes the beds.' Mark does not like 'being picked
on' by other children. 'John is a bit of a worrier, he worries
about his school and his biscuit money ... he gets sick if he's
had a fight, he hates fights, it's usually the other children who
start it, they steal things from him.'

What sort of games does he mostly play with other children?
(Prompt: rough and tumble/building/making/imaginative games.) Two-
thirds of the mothers merely picked one of the prompted responses,
mostly rough and tumble, sometimes adding to it: 'He likes kicking
a ball.' 'He best likes climbing in empty houses down the terrace,
he likes to think he is Tarzan.' 'He plays in an old car.' But
others enlarged: 'He's forever getting old boxes and playing with
them in the yard.' 'He makes dens.'

One-third of the mothers described more varied patterns: 'Charlie
likes looking at books, he plays with the ring board, ludo, and he
likes playing hide and seek. He will play football - I don't like
balls, not in the road, it's all right in the park. I do away with
them, Father burns the balls. He makes dens, he plays with bucket
and shovel, he plays at shops with the girls. If I give them 10p.
for a party they put chairs out and have a little feast, they play
Mothers and Fathers.' Others mentioned cowboys and Indians, doctors
and nurses. 'He covered a doll with red spots and said she had
measles.' Others merely said, 'No specific games I've heard them
playing - yes, rough and tumble - never constructive or imaginative
games.' Imaginative games as a first preference was mentioned by
only four families, in contrast to nearly one-third of Nottingham
non-skilled families.

*In large families it is not always possible to manage pocket
money for everyone. Does he have any? And what happens when he
gets it?* It is common practice to send the child off to school with
a small amount of money every day, if it can be afforded, and this
is referred to as 'biscuit money'. 'They get some for school, but
not every day.' 'Sometimes if their Dad's got it, he'll get it -
sometimes they'll have it two to three times a week.' 'He gets 3p.

morning and evening - they worry the life out of you, I don't like to see them eating a lot of sweets.' All but five mothers gave an account of this kind. The pattern is that if there is money it will be given and divided evenly between the children before they are off to school. 'Biscuit money' is often supplemented at the weekend by another sum, sometimes earmarked for special things, such as a comic, or to go to the pictures on Saturday morning. In most families mother doles out 'biscuit money' and father gives something at the weekend. Sometimes the children get money from their brothers and sisters at work, sometimes this is a regular weekly gift. Other relatives give money, uncles, aunts and grandparents were mentioned, but these gifts are not frequent or regular. One father said 'My mates give him something when they call.' One boy earns a few pence regularly 'for errands for a lady in the yard'. Five families do not distribute money at all, but mother usually provides sweets, and occasionally gives the children a coin.

The money is usually spent immediately on sweets, pop, biscuits, ice lollies, and bubble gum. But some spend it otherwise: One boy 'has a mania for cars', one buys marbles, one 'may buy a ball', four boys will save. Charlie's mother: 'He likes jingling the money in his pocket. He tries to hold on to his money. If he buys sweets he'll always hand them round. I've got a penny bank, near Christmas they all give me their pennies and we save it for all of them.' 'Chris has had relations give him money, he saved that for Christmas, he's managed as much, he's not greedy like some. If something is really necessary - like socks - he'll go and buy them.' Eamon who attends a Catholic school, 'saves his money sometimes and gives it to teacher for black babies'.

At this age children often like to collect things. Does he do this with things like bubble-gum cards, badges, or toy cars? Does he spend his own money on this? Seven-year-old boys in Nottingham show a descending scale of collectors, with 85 per cent in classes I and II and 74 per cent in class V. The percentage in this study was 71. The objects mentioned by their mothers are cards in tea or bubble-gum packets, milk-bottle tops, match sticks or boxes, old tin cans, little bits of string and 'tat'. Some collect marbles: 'He likes marbles, the dustman usually brings him some'; some collect toy cars: 'He likes collecting cars - he still tries but he has them pinched, he had about twelve.' Charlie's mother: 'He collects cards from tea bags, he takes the stamps off the tea bags and sticks them on to a card and counts them; he likes marleys, he loves pennies and tuppences, he collects milk-bottle money, he likes me to keep them for him (1p. was paid for sterilized milk bottles), he hides pop bottles so that he gets the money and the others don't.'

Has he a special place of his own where he can keep his own things? Only one mother said 'Chris has a box, but they are usually all over the place', another three boys have a special place in which they play, in the attic or in one of the bedrooms. Otherwise there was no provision of a cupboard or shelf that the children could consider their own. It reflects housing conditions, and the absence of toys; it also reveals a striking contrast between our sample and the Nottingham study. Among professional families (class I and II), 57 percent make such provisions, and even among non-skilled families a quarter are able to do so.

Chapter 8

Some children like to have their own special toys and some think that all the toys should belong to the whole family. How do you feel about that? The replies, as can be seen in Table 8.1, indicate that the mothers in our sample have developed an attitude which is very different from the mothers who took part in the Nottingham study. Three quarters of the boys have no personal possessions except the odds and ends they collect: 'The toys belong to the whole family - they have only one ball, I don't let them say "It's mine".' 'I think communal is best, when they only have one thing they seem to fight over it. If I get anything I get it for them all.' 'In this home all toys are shared.' 'If they have their own special toys I think they get spoiled.' 'I've tried individual toys, it doesn't work, they all squabble over it.' The few families who do try to encourage individual ownership of toys encouraged sharing too: 'He has his own, but he does play with other toys too.' 'At Christmas they all get their own toys (supplied mainly by Family Service Unit), but they all play with them - Sammy likes things to be his own.'

TABLE 8.1 Ownership of toys (percentages)

Class:	Nottingham I&II	IIIn-m	IIIm	IV	V	All	Socially handicapped families
Separate	52	47	47	51	56	49	14 ⎫
Some of each	39	39	36	21	21	33	11 ⎬ 25
Communal	9	14	17	28	23	18	75 ⎭

Twenty-nine boys aged 10-11 years

This age group of boys was described by their mothers in terms not unlike the younger boys. Twenty-one mothers emphatically stated that their sons were 'outdoor types': 'He's never in, he'll come in when his favourite programme is on.' 'He only comes in for cowboys on the telly.' 'You never see him - he's the last one in for meals.' Five mothers described their sons as liking both, and two said the boys were 'indoor types': 'He doesn't play with anyone - he's more indoors.' 'He doesn't go out much unless there is a football match at school.'

Does he enjoy being all on his own, does he like his own company? How does he get on with other children - does he make friends easily? Fifteen boys were described as making friends easily, eleven as intermediate, and three found getting on with other children difficult and preferred their own company. Twelve of the fifteen gregarious boys belong to a group who 'always go around together'. Mothers, when asked, sometimes appeared a little uneasy: 'They don't go in a gang causing trouble.' 'They are just friends, three or four of them.' 'They are always a crowd, when they are playing football.'

The feelings of parents in relation to undesirable influences were further explored: *Are you happy about his friendships or do you think some of them have a bad effect on him? Have you done anything about that?* Thirteen mothers said they were satisfied: 'I'm quite happy - Raymond is a good judge of character, he won't be led into trouble.' 'They are pretty well all the same age - Ronald's a lad that'll do what *he* wants, whether anyone tells him or not.' Three mothers said their sons had no friends, and another one could not judge. But twelve mothers expressed anxiety about boys in the neighbourhood. Their approach to the problem varies. Some parents try to prevent further contact by keeping the boys indoors or away from undesirable company. 'I don't let him play with the gangster kids around here - he makes friends all right in school. At home he's mostly on his own.' 'He plays more with his brothers, I keep him that way, I don't like the company ... yes, *he* does want others.' 'He's a good lad, but over the years he's been under a bad influence, he's tended to pick up bad ways - stealing. He still steals, I took him to the probation officer three weeks ago, he's only young, but I'd like him to be straightened out now, get it cut in the bud before it's too late. I think it's other lads that cause it, if they dare him. If I could afford it he'd be off swimming or playing football.'

Other parents have different methods: A father: 'We try to keep him away from them because he's easily influenced, but once they are out, it's impossible to keep an eye on them.' A mother who felt very uneasy about her son's friendship with a boy who was eventually sent to approved school had tried 'to turn him in the opposite direction' but had failed; her son's present friend, however, was 'not one of the ones who'd lead him astray'. This boy in fact became delinquent aged twelve. A father described his son: 'He's got a friend what's a bit of a rogue, if he won't go with them stealing they call him "toffee nose" ... I disapprove of him going with Allan, he's strong enough to come away if he knows he shouldn't. We wouldn't stop him - we warned him, and he knows.' This boy was not caught by age thirteen. A mother: 'He's got one I don't like ... If I think a child goes out pinching I wouldn't let him play with him, I'd explain to him why. I always explain what I'm doing, maybe that's why I have a good relationship with him.' She seemed to be successful; other mothers were less good at explaining: 'I don't like him playing with them, but what can you do - they're bad company, they get in trouble with the police. I've tried to stop him, he won't listen to me.' 'There are a few up here I don't like him to mix with, but I can't really stop him. Once he's out he can do what he likes - perhaps when he's older I can talk to him.' 'One boy had a bad effect on him, he led him into trouble, they used to play truant together. I've slapped him, Father has slapped him, I've put him to bed with no tea.'

Is he one of these children who is always in hot water, or does he manage to keep out of mischief mostly? Thirteen boys were described as always being in trouble: 'He's always in trouble with the neighbours' kids - they keep picking on him.' 'He won't go outside but someone thumps him.' The rest manage to keep out of trouble usually.

Sometimes children at this age go through a phase of taking things that do not belong to them - have you had that at all? What did you do? Sixteen mothers denied such happenings, one of them concealing the fact that her 10 year old had just appeared in court for stealing. Thirteen mothers gave accounts of various delinquencies. Four boys were said to be pinching things at home only: 'Only with food, that's cakes and such, I count them, I find out.' 'I tell him, if you want anything ask, don't take, I'll give it you if I can give it.' Others admitted more serious trouble: 'He took my purse, and he's also stolen from the shops. He's never robbed Father, only me – maybe he thought I had more. When I first found out I didn't accuse them straight away, I used to ask them, they all said "no", I tried to catch them but couldn't.' 'I have had trouble, me and my husband – but I haven't had any trouble for some time now. I went to the school. He wouldn't tell me the truth, so I took him to the police, but I'm afraid they couldn't help, in the end he told me the truth.' 'He tried to break into an empty house once, he thought it was empty – nothing apart from that, he returned the things to the lady. They called the police but nothing was done about it, he was too young.' 'He took metal from the tat yard – I didn't know till the police came and knocked on the door.' Of these thirteen boys, five had already had a caution or had been found guilty of an offence in the juvenile court. By age 13 another five were cautioned or found guilty, as reported in chapter 6.

When asked if the boys ever got into trouble with neighbours, nine boys were mentioned: 'Neighbours fetched the police to him for taking the milk, they accused him of taking their milk.' 'He was climbing on a toilet roof, shouting and going up the far end of the yard.' Others had been caught climbing into the park, or into the football ground. In all, twenty of the twenty-nine 10-11 year old boys were reported by their parents as having got themselves into trouble.

When he has to amuse himself is he a busy sort of a person or does he easily get bored? Twelve boys were described as generally busy when indoors, nine as varying, eight as bored. The activities of the busy boys: 'He always finds something to make, with paper mostly, he makes models.' 'He is interested in modelling kits, he's got two, one for his birthday and one for Christmas.' Others mentioned reading, writing, drawing, playing cards, playing with marbles or match sticks. The seventeen boys who sometimes or often get bored: 'He can't keep still – even watching TV he's up and down like a Jack in the box.' 'If he's got something to play with, he'll do it for five minutes and then he's off to see what the others have got.' 'He gets bored and looks for mischief.' Some mothers said the boys liked to be outdoors, but did not know what to do with themselves if they had to stay indoors. Asked if the boys had any special interests, the constraints of circumstances became evident: 'He likes modelling, if I can afford it I get these Airfix things that you glue together.' 'Give him a note book and pen, he'll sit and draw for hours.' A boy who was present during the interview told us: 'I play chess in school, in class and after at play time.' Another boy told us: 'I like reading, at present I'm reading "Just So" by Kipling.' We saw this boy read to younger siblings by the light of a candle when the electricity supply had been cut off for non-payment of a bill. Six boys go swimming, one boy takes himself to the museums in town. Eight belong to an out-of-school club or church group.

Earlier in this chapter the feelings of mothers are reported about the cost of toys and their inability to provide them. Like the 7 year olds, the older boys get 'biscuit money' on going off to school, and some of the boys get money at the weekend from their fathers or sometimes from a brother or sister at work. Three-quarters of the boys have opportunities to earn a little, mainly by 'running errands' for mother or for neighbours. One of the boys, a very bad school attender, was earning some 15-20p. daily by helping a scrap collector. Another boy keeps an old man's little patch of garden tidy for which he gets about 50p. a week. One boy had a paper round for the newsagent, but he stole a wrist watch when he was invited into a house for a cup of tea. His mother did not allow him to continue this work. Some boys get paid by their mothers for the shopping they do, but some mothers did not consider this right: 'They have to learn to do things without being paid.' The money is usually spent at once, mostly on sweets, ice lollies or 'food'. Some boys buy other things: 'He might bring home a bunch of flowers.' 'He'll buy me and the kids something, he's quite generous.' 'They don't buy a lot, they say, if Father is still out of work by Christmas they will help with food presents.' 'If I'm hard up he'll come and give it to me.'

Does he ever come and say 'so-and-so is allowed to do something - why can't I?' Does he ever complain that another child has something - more pocket money or something like that? Thirteen mothers said the boys grumbled or commented now and then about things that other boys possess or can do; sixteen gave answers indicating that the boys are resigned to their state: 'He knows the way things have been going with Dad's work and how badly off for money we are.' 'He realizes I can't do it.' Grumbles are countered by mothers: 'His mother has only got one child, not all you lot.' 'I tell him about prices going up - we are a bigger family and can't afford it.' But three mothers 'give him a clip', 'belt him' or 'hit him' if it gets too much, and one said 'Father would soon stop him.' Some of the mothers related the question to restriction of freedom of movement, not to lack of money: 'No good giving them too much freedom.' 'I tell him Robin is four years older.' 'Some parents are slack - they'll be sorry and the children control their parents.' 'I tell him if that little boy is allowed to do that then his Mum can't be worried about him.' 'I tell him what's right and what's wrong.' If these boys do feel deprived and envious, most of them do not express such feelings frequently to their mothers.

FAMILY ACTIVITIES

What about the things you might do as a family, or several of you together? For instance do you manage a family outing somewhere? Do you sit down and play a card game or a board game together? What about holidays - have you managed to have one with the children? 'We used to go on outings, but it's a heck of a lot of trouble to take them all out.' 'We went to Blackpool last Saturday - first time we've been on an outing for five years.' 'We had a day outing to Weston last summer.' Holidays with overnight stay elsewhere are an unknown pleasure to almost all families. Some families take the younger children to a near-by park or the reservoir, but more than half do not. One mother told us: 'A couple of years back we used to go to Stafford prison to see his Dad - see him and go in the big park there, he liked that.' A father: 'If I had more time I would -

but I work weekends - I'm tied on maintenance, I've had to do seven days a week. If I had more leisure time I'd take them swimming and to football matches. At Whitsun I took them to the Botanical Gardens.' 'He's never asked to go - I expect we would if he asked. He's been to Bingo, he came up the club with us, we don't go out much.' 'We took them Thursday to see "Alice in Wonderland". It was Father's idea, but it was more trouble with them than I ever had, the boys fidgeting, Lisa went to sleep, Brian was niggling, only the girls were okay.' Games are played as a family entertainment in twenty families; some of these added qualifications: 'In the winter we might.' 'At Christmas.' 'We used to, but lately the telly has been more important.' 'I've never had the patience, Father sometimes plays with them.' The games mentioned were ludo, snakes and ladders, draughts and card games.

Mothers of the 10-11 year old boys were asked: *Do you find you talk a lot to him? Or are you not really interested in the same things?* Thirteen mothers said they had no common interests: 'I don't bother much with him.' 'I don't think we do talk much - more with the girls than the boys.' 'When he wants something he'll talk to you, clothes and that - "can I have them?" We don't share any interests.' 'Ronald isn't one you can talk to, he only comes when he's in trouble.' 'He doesn't see eye to eye with me, he's all for his Dad.' 'Depends on his mood.' Seven mothers said they talk but don't have a lot to say to one another: 'I talk to him as much as to the others - about school and such, I answer his questions as best I can.' 'Not a lot, when we do he doesn't say much at all to be honest, he says what he did at school.' 'He's quiet, he'd sooner look at a book. I like to talk to him to try to get a bit closer to him, he doesn't talk easily.' Nine mothers said they talked a lot with their sons: 'It was a case of having to talk to him a lot, he was going through an emotional upset, he thought we didn't care for him.' 'Tony and I talk a lot, he's very understanding; he wants to know about his real father, he thinks he died at sea. "Is he watching over me?" I say "Yes, he is".' 'We do a lot - and when I fall asleep in the chair he'll put an "Out of order" sign on my head.' 'I can talk to him more than I can to his father, he listens and asks questions, I don't feel any shyness with him, freer in some ways than with Father, especially at present.' 'We talk about all sorts of things, we talk about films on TV, he brings things in and asks me what they are.'

What about his daddy? Do they talk a lot together? Seven mothers said that their husbands had little in common with the boy: 'He doesn't talk with Tony, they have nothing in common, he's said once or twice "You are not mine anyway".' 'No shared interests - Father has been moody these past few weeks, Keith likes to come to me.' 'They don't get on, Father hates Michael, probably because he looks like him, he's always been like that since he began to grow up.' But the majority of mothers, nineteen in number, spoke warmly about father-son relationships: 'He likes his Dad, they talk about men's things, football and cricket and such like.' 'Pigeons and animals.' 'Football and swimming, they are both for being outdoors.' 'They get on all right, both like fishing.' 'They are both as bad as one another - making a model car now.' 'He holds conversations with them for hours.' A father: 'He talks a lot to me in the garden, and when I'm painting he likes to have a go and learn' (house painting and decorating). Three mothers, temporarily alone, were not asked. Two-thirds of the boys thus seem to have good relationships with their fathers; with two exceptions the same boys also have good relations with their mothers. Seven boys communicate little with either parent.

A series of questions was asked to ascertain special interests that the 6 year old or the 10 year old boy might share with his father, or with his mother. But what special interests did the parents have? An index of activities in the home was constructed recording parental hobbies (including reading and knitting, but excluding television). The possession of a radio, camera or other instrument for leisure use, books or the regular reading of a newspaper were also recorded. Over half the families engaged in no more than one of such activities, fourteen in two, and eleven in three or more. These families possessed a number (not more than ten to twenty) of hard-cover books, some kept tropical fish, some had birds in a cage, or bred pigeons, some grew flowers and vegetables, some tinkered with old radio and television sets, or did carpentry around the house and made toys for their children. One father made a wendy house with instructions given by two pre-school children who attended our play sessions. One man, who had been employed as an excavator driver, had acquired an interest in historic objects he had dug up. He photographed his finds. Prolonged spinal trouble and dependence on sickness benefit curtailed these activities. A number of fathers went fishing. One mother tried her hand at writing short stories.

In a milieu as devoid of resources as this it may appear over-meticulous to investigate whether the number of leisure-time activities was related to income; it was done and, indeed, a relationship was found even within so narrow a band of incomes. Of those whose incomes are below the poverty line, over two-thirds had no more than one hobby, whereas among the seven families above the poverty line five engaged in two or more hobbies, which suggests that even such simple activities as knitting or the regular reading of a newspaper are outside the reach of those on a poverty income.

Families with more hobbies provide more opportunities for the 10-11 year old boys to share parental interests; some take the boys fishing to the reservoir, some let them help with decorating, or gardening, or father and sons tinker with old radio sets. One father was busy making a chart with his son for the World Cup. 'If Father is outside doing Chris's motorbike or something electrical or mechanical, he's there too.' In other families, where there are few parental activities, mothers nevertheless indicated a readiness of their husbands to spend time with the boy, or boys: 'They go and get wood for the fire.' 'He likes to have a rumpus on the floor with them.' But many mothers stressed that 'Father doesn't favour one above the others', or 'doesn't single one of them out'. In forty families (61 per cent) father was reported as liking to do things with his sons. Taking both age groups together, fathers' participation was significantly related to the number of parental hobbies or interests, as shown in Table 8.2.

TABLE 8.2 Hobbies and father participation (boys aged 6-7 and 10-11)

	Participation			Total	
	High	Fair	None		
Hobbies				N	%
0-1	6	14	10	30	54
2 or more	13	7	6	26	46
Total	19 (34%)	21 (37%)	16 (29%)	56	100

(Chi-squared = 5.59 with 2 df, p < 0.05 on one-tailed test.)

Family activities include the help that many parents give their children in learning to read and write. The mothers of the younger age group of boys were asked: *Have you tried to help him with his reading at all? Do you or his daddy ever help him with other things like sums or writing or any other school work?* The answers, given by our sample and that in Nottingham, are shown in Table 8.3.

TABLE 8.3 Parental help with school work (boys aged 6-7) (percentages)

	Nottingham						Socially handicapped families
	I&II	IIIn-m	III m	IV	V	All	
Regular or occasional	75	88	80	79	65	79	36
No help	25	12	20	21	35	21	64

It appears that parental involvement in the acquisition of the basic educational skills is not related to social class in the Nottingham sample, with the exception of a drop in class V. The proportion of children in our sample who get help, a little over one-third, is only half that reported of nonskilled families in Nottingham.

Ten parents help their sons: 'Well, I've read to him and I've asked him to read to me, and I've asked him words.' 'He's never had a book from school, but I help him with his name and address and how to write it.' 'Chris gets homework.... He came in and he had his two-times table. I wrote them all down, I wrote it all down without the answers, and I asked him.' 'We often write out sums for him, and he enjoys copying writing.' 'I help him with the money lessons.' The reasons given by eighteen mothers who do not help, and whose husbands do not help, are many: 'He's never brought any work home - anyway I'm not good at them.' 'I'm as daft as them.' 'He's not interested.' 'We've tried, but he doesn't co-operate.' 'He never brings home any sums. Father has told him before to ask for sums to keep him occupied at night, but Miss won't give them any.' 'We found that Brian gets hold of things and rips them up - we asked the Head not to send books home with them, they never looked at them when they brought them home.' The 7 year old son of one mother, who told us that neither she nor her husband had helped the boy, is a fluent reader.

We asked: *Does he ever come home and start doing something he has been doing at school?* Twelve mothers mentioned drawing: 'He asks for a biro and paper and he draws and says "That's what I've done in school".' Others talked about the books their children brought home; others about their efforts to write: 'He does writing if he does anything, he's happy with a pen or pencil and copies for bloody hours.' Eight mothers could not think of any activities their sons engaged on at home that might have been learnt at school. Later in the interview, returning to the same topic, we asked *Do you get him comics regularly?* Nineteen mothers said that the boys usually looked at the comics which older siblings had bought, but the other nine were not sure that their sons appreciated comics: 'He might have a comic, but he can't concentrate, he gets fed up.' 'He does look at comics and books – the mood's got to hit him – he prefers to be out.' 'No, they are not interested in stories, they fidget all the while.'

So far we have not said anything about involvement of fathers in the management of their pre-school children. Mothers were asked: *How much does your husband have to do with him? Does he play with him a lot? Bath him? Dress or undress him? Read to him or tell him stories? Take him out without you? Look after him while you are out?* Answers were grouped in three categories: 'high', 'fair', and 'no participation'. The distribution, as shown in Table 8.4, shows considerably less involvement by fathers than in non-skilled families in the Newson study; only just over one-third had a 'high' rating, and one-fifth did not undertake any activities we asked about. Two factors may partly explain the lesser degree of involvement by fathers. We have earlier discussed daily routine; the typical pattern is one of outdoor activities for all children commencing at an early age. In chapter 9 child-minding by older siblings is discussed, which again is very common in large families who live in overcrowded conditions. Thus father involvement may not be required to the same extent as it may be in other families. Nevertheless, nineteen fathers shared with their wives many of the daily activities concerning the 3-4 year old.

TABLE 8.4 Father participation (4 year olds) (percentages)

Class	Nottingham						Socially handicapped families
	I&II	IIIn-m	IIIm	IV	V	All	
Participation:							
High	64	59	48	44	49	51	34
Fair	32	36	42	42	41	40	45
Little or none	4	5	10	14	10	9	21

Traditionally working-class husbands were not expected to help with the care of young children. In the words of Hoggart (1957,

p. 38) 'When all's said and done, most things about a house are women's work: "Oh, that's not a man's job" a woman will say, and would not want him to do too much of that kind of thing for fear he is thought womanish.' Men come home tired, often late; if they are out of work traditional attitudes do not change overnight. That they are changing was already indicated by Hoggart in the mid-1950s, and is borne out by the facts shown in Table 8.4. There are other factors. Husband-wife relationships are important, and so are other stress situations which affect personal functioning; all of them of considerable significance in a sample of families who are disadvantaged. We have earlier discussed the association of stress and home atmosphere (chapter 7). It is reasonable to assume that parents who are relatively less preoccupied with their own worries and less burdened by stress, such as the care of an invalid spouse or child, or tense marital relations, will be more ready to join in the activities of their children. We tested the association of the atmosphere of the home and parental participation and found that this assumption may well be correct, as shown in Table 8.5.

TABLE 8.5 Parental participation and home atmosphere

	Participation			
	High	Fair	None	Total
Mothers (3-4 year olds)				
'Happy' home	10	5	4	19
'Unhappy' home	1	4	7	12
	(Chi-squared = 6.7, $p < 0.05$.)			
Fathers (3-4 year olds)				
'Happy' home	8	11	0	19
'Unhappy' home	3	3	6	12
	(Chi-squared = 12.0, $p < 0.01$.)			
Fathers (6-7 and 10-11 year olds)				
'Happy' home	9	9	1	19
'Unhappy' home	0	4	8	12
	(Chi-squared = 15.5 with 2 df, $p < 0.001$.)			

(Significances have been checked by the Fisher 'exact' test, only the 'high' and 'none' categories being used. The results are 0.012, 0.007, and 0.0002 respectively.)

Mother's participation in the activities of the 3-4 year old was discussed earlier in this chapter; the definition of participation is a less stringent one than that applied by the Newsons and includes any form of play with the child. However, to include the Newson definition, only those mothers who scored as participating wholeheartedly in play at the child's own level have been categorized as having a 'high' degree of participation. As shown in Table 8.5, ten mothers in the 'happy' homes, but only one mother in the 'unhappy' homes achieve this. The association of the two variables is even stronger in the case of fathers, both for the younger and the older age groups.

In relating what we know about home circumstances in parental practices a picture begins to emerge of typical families. In this chapter an account has been given of the child's world, and it was shown that the children, in spite of their shared experiences of poverty, are engaged in a fair range of activities and are exposed to a variety of relationships with their parents. In summing up the findings we have divided the families into three groups on the basis of their school boys' general activities. The younger age group of boys was rated on the following items: he draws or paints at home often (score 2); he collects things (score 2); he has private possessions (score 2); he has a special place to play (score 1); he plays constructive, imaginative games (score 2); he is happy to sit still when he has something to do (score 1); he is always busy (score 1). The older age group was rated on the following items: he has at least one special interest (score 1); or several interests (score 2); he is a member of an organization (score 2); he is always busy (score 1); he often plays games at the table with other members of the family (score 2); he goes on family outings (score 1); or family holidays (score 2). As expected, although the maximum score for the younger age group is eleven, and for the older age group ten, none reached this, and our definition of 'intensive activities' is based on sixteen boys who scored five or more. At the other extreme, 'deficient activities' singles out seventeen boys who scored two or less. An intermediate group of twenty-three boys scored three to four. We have not been able to establish any associations of the three categories of activities with family characteristics; there is a strong trend for fathers' participation to be higher in families whose boys have more intensive patterns of activities (chi-squared = 6.54 with 4 df, $p < 0.2$), but it fails to reach significance. The index of activities is not a simple indicator of provisions made by parents, it also gives information about the boys' personality attributes as described by their mothers. It is well to remember that the range of activities that the children in our sample enjoy is a very limited one.

Chapter 9

ACHIEVEMENT, SELF-RELIANCE AND RESPONSIBILITY

INDEPENDENCE TRAINING

Fifty-six children aged 3-4 years

The Newsons point out that it is a characteristic of our culture at this time 'that children of pre-school age are required to do hardly anything which could be called work, or which is either necessary or even useful to family or community'. But there are two exceptions: 'One is tidying up, the other is the running of errands' (1968, pp. 67-9). Tidying up concerns mostly the child's own things; taking messages, however, and shopping for mother are tasks by which the child can make a contribution to the household. In the Nottingham study 62 per cent run errands for their mothers, in our sample 66 per cent do. These are tasks that the 4 year old will find easier than the 3 year old, but there was no substantial difference between thirty-four older and twenty-one younger children. Shopping by himself involves greater social and linguistic skills. The question asked was *Does he ever go into a shop or to an ice-cream van on his own (while you wait outside?)* Seventy-two per cent of Nottingham's middle-class and 81 per cent of working-class children shopped on their own, in our study 71 per cent did (76 per cent of 4 year olds and 64 per cent of 3 year olds). While mothers, on the whole, encouraged their children to shop by themselves, mostly to get their own sweets, occasionally to get something for the family, nearly 30 per cent maintained the child was too young, the road was too busy, or simply stated that 'He won't go inside even while I wait outside.' Three mothers referred to language difficulties: 'He can't talk very well, so I don't send him.' 'Not everyone finds her easy to understand.' Many have reservations about the ice-cream van: 'He'll go to the shop when he gets pennies, he doesn't need to cross the road - but not to the ice-cream van, it might be on the far side of the road and he'd have to dash across.' 'He hung on to the van once and was nearly dragged away.' 'She doesn't go to the ice-cream van, none of them do - I've seen two children injured and one killed, only the big ones go.'

Independence is also achieved by a young child in dressing and undressing himself, tidying up his own things, and going to the lavatory alone. The Nottingham enquiry showed 59 per cent of children to be fairly independent; there was no significant class difference in overall achievement. Our sample showed that the young child was considerably less able to look after himself. For comparison purposes the group was divided into 4 and 3 year olds; only about a third of 4 year olds and fewer 3 year olds had achieved a degree of independence, as shown in Table 9.1.

TABLE 9.1 General independence of 3-4 year olds

	Fairly independent No. %		Rather dependent No. %		Total number
Nottingham study					
4 year olds	59		41		700
Socially-handicapped families					
4 year olds	12	(35)	22	(65)	34
3 year olds	6	(27)	16	(73)	22

(The difference between the 4 year olds in the two studies is significant: chi-squared = 7.5, 1 df, $p < 0.01$.)

The comments given by mothers indicated that, in the majority of families, maternal expectations are low: 'He doesn't do that, he's pig-headed - I keep telling him. He says, I don't want to, I don't have to, and so I do it myself - he's got his father's habits.' 'Sometimes she's tidy if she's in a good mood - she says "oh blast" and throws her clothes on the floor.' 'You must be joking, none of them are tidy, he doesn't clear his things up, all of them are like that.' 'In hospital the nurses made him tidy up, at home he leaves them lying around.' 'She drops her clothes and she says to me "put them up for me, Mum".' Only eight mothers insist on tidiness: 'She leaves her things around, if I ask her she'll do it herself; when she sees the others doing it, she'll tidy them, she doesn't need a lot of help, she's a busy little thing.' 'I keep reminding her, I have to keep on till she does it.'

Maternal low expectations are coupled with the practice of delegating certain mothering activities to older children, who may be a great help to their mothers but who are not likely to be of much assistance in the training of the young child to achieve independence. In answer to several questions concerning the sharing of bedrooms, bed times and help given by others in putting young children to bed, thirty-four mothers (61 per cent) said they were in full charge, twenty-two mothers (29 per cent) named older children who helped a great deal, and ten mothers (18 per cent) said fathers helped a great deal. We have selected some quotations which illustrate these aspects and the circumstances that form the back-cloth to daily life: 'The big ones manage themselves, and they will also help the little ones. In the big room there are four boys and Christine, two boys are in the box room, and the little one sleeps with us.'

'Tony (aged 10) looks after them all, he's very good, he helps a lot bringing Mandy down, dressing and undressing them and playing with them.' 'All the young ones go together, sometimes Seamus helps, he stops up with me. The four boys sleep together, the three girls with us.' 'When I'm in, I do it myself; if you ask Father you never hear the end of it - "it's your job and all that". They are all five in one room, I have a bed in the attic, but Stephen says he won't go up there, it's too cold.' 'They all go together, sometimes Bill goes later - Bill carries them if they are sleepy.'

Bedtimes vary greatly, the times given by mothers ranged from six to nine pm. The very youngest often stay up until the parents go to bed. Frequently they share their parents' bedroom, and sometimes the parents' bed. The simplest method of achieving an undisturbed night is to let the young child play until he falls asleep: 'Kevin stays up a bit, we let him play till he drops.' This bedtime routine, involving a reversal of the usual age-related order, was encountered in fourteen families. Many of the teenage children and some of the ten year olds had no fixed bedtime. They go to bed when their parents do or 'when they feel like it', or 'are ready'. Under conditions of severe bedroom overcrowding where different age groups share bedrooms or even beds it is difficult to operate a system of fixed bedtimes related to age.

Twenty-eight boys aged 6-7 years

At this age, the final year at infant school, the boys spend much time out of doors. How mothers feel about their sons' increasing freedom of movement is a vital aspect of child-rearing; the demands made by mothers and the sanctions used to enforce them will affect the boys' growing self-reliance, and will have an impact on the boys' activity patterns and experiences. The information contained in Table 9.2 shows the patterns of experience of Nottingham boys, arranged by social class, which is based on a series of questions beginning with *Could you tell me what sort of things he does on his own?*

TABLE 9.2 Activities undertaken by the 7 year old on his own (percentages)

Class:	Nottingham						Socially handicapped families
	I&II	IIIn-m	IIIm	IV	V	All	
Shops	88	86	88	90	95	88	93
busy roads	36	38	41	43	53	41	75
bus	23	16	10	12	17	13	(4)*
park	36	52	48	59	68	50	29
baths	3	5	3	4	4	3	(4)*
pictures	7	10	11	11	18	11	(7)*
elsewhere	7	19	12	14	12	13	(4)*

* The number of boys, one or two respectively, is too small to be expressed in per cent terms.

The problem, how much independence to give a 7 year old boy, should of course be related to environment. Traffic in the inner city may make shopping much more hazardous than in the suburbs; private gardens make the visit to a park superfluous; and the vicinity of a park in the inner city may make access possible for a limited number of boys only. The most widely-achieved test of self-reliance in our group is shopping, all but two boys go shopping alone. The two exceptions are restricted for very different reasons. 'Eamon likes to go to the park, he's always at me to get out of the yard, but I won't let him... I have to watch him with his eye - I just give a shout, they don't go far away.' (This boy has an eye defect.) There are other reasons for keeping this boy and his siblings in the yard, a neighbour's child was drowned in a nearby reservoir. In contrast, Sam is kept indoors or in the back yard by an overprotective mother who does not like any of her children to leave her. We were unable to persuade her to let the 4 year old attend the play sessions. 'I don't mind - they are company for me, if I go out they come with me, we go out for chips down the road.'

Crossing busy roads is achieved by three-quarters of the 6-7 year olds. Mothers were not always happy about this, but they were in a quandary. They could not always arrange company of an older child on the way to school and back, but only ten mothers took and fetched them. The restrictions on swimming, the cinema, bus journeys and 'going elsewhere' were largely imposed for financial reasons. Two further activities, measuring degree of independence, showed up sharply the relative deprivation of the boys in our sample. In the Nottingham study, membership of a club or an out of school organization was 12 per cent for semi-skilled and 4 per cent for non-skilled families' boys. Only two boys in our sample attended clubs, run by social workers. The experience of visiting a friend or relations overnight was known to 18 per cent of Nottingham's non-skilled families' boys, but only to four boys in our sample. (This does not include the fostering of these children during periods when their mothers are laid up.) In summary, only one-third of the boys in our sample achieved three or four independent activities, of which shopping and crossing busy roads are universal, in contrast to half of Nottingham's working-class boys.

A further series of questions was concerned with mothers' reliance on the boys' responsible behaviour when out of sight. *Do you have rules about telling you where he is going before he goes out?* Half the mothers were quite firm about rules: 'Father is strict - he gets on my nerves sometimes wanting to keep them in all the time, he always seems to be concerned with what they are doing.' 'I don't let him go anywhere without telling me.' 'I like him to let me know - even when he's going with other boys, what he will do.' The other half was more vague: 'I normally know where he's gone.' 'He doesn't go out too often - he will tell me where he is going, but I don't make him.' 'He usually tells me where he goes - that's not to say he doesn't wander off from there - once the police had to look for the three of them.' 'I'd like him to tell, but he just goes.' 'They just go out, they never tell me.'

Do you have any rules about coming straight home from school? Ten mothers usually escort the boys. Most of the others are expected home straight from school, many are in the company of older siblings. Only two mothers were quite vague: 'He plays out after school, though I'd like him to come straight home.' 'He should come straight home, their Dad is always telling them to.' After tea, *Does he play or roam around in the streets at all?* Two-thirds allow the boys to roam freely; many mothers pointed out 'He plays on the streets because there is nowhere else.' *Can you always find him?* 'No, not all the time, the other night it was nine before he came back.' 'I can never find him, I have to shout in the streets for him.' Nine mothers place restrictions on the boys' movements: 'Joseph never goes anywhere, but to the shops, on his own. I don't let any of them, it's not so much the road as the people, in the past twelve months there were three or four who attempted to take children away.' 'He doesn't play in the streets, I keep him in the back.' A smaller percentage of boys are allowed to play in the streets or roam than any boys in Nottingham, as shown in Table 9.3. This must be a reflection of the perceived hazards of environment in the deprived areas in which our study was located. The percentage of mothers who cannot find the boy who is allowed to roam is high in contrast, but roughly in line with non-skilled mothers in Nottingham.

TABLE 9.3 Roaming and finding of 6-7 year olds (percentages)

Class:	Nottingham						Socially handicapped families
	I&II	IIIn-m	IIIm	IV	V	Total	
Roams	75	78	83	75	89	81	68
Often can't find	13	12	16	28	25	18	29

What is the relationship of having strict rules and the fact that mothers admit they cannot always find their sons? Not all the mothers who maintain they know where the boy is rely on rules, and quite a number who say they have rules admit that sometimes they cannot find their sons. Some mothers who can always locate their sons do not operate a system of rules, they rely on keeping a close eye on the boys' movements (Table 9.4).

TABLE 9.4 Rules about street play and locating the 6-7 year olds

	Rules	'He generally tells'	No rules	Total	%
Mother					
- always finds	7	2	1	10	(36)
- sometimes not	6	1	3	10	(36)
- often not	1	2	5	8	(29)
Total	14 (50%)	5 (18%)	9 (32%)	28	(100)

Chapter 9

Twenty-nine boys aged 10-11 years

The boys in this group are mostly in their final year at junior school and at a developmental stage which is accompanied by expectations of increased personal independence and responsibility. The way the boy is seen at home could be affected by his ordinal position; the eldest could be expected to assume a degree of independence that a boy lower down in the birth order would be considered too young for. In fact, only four of the twenty-nine boys are first born and none are the youngest, as all families had a pre-school child. There is a wide range of ordinal position.

Mothers were asked a question similar to that which mothers of the younger boys had been asked: *We would like to know what sort of things children of this age do on their own?* The replies showed that nineteen boys (66 per cent) use buses, seventeen boys (59 per cent) go to the park or elsewhere on their own, but only nine (31 per cent) go to the shops in the city centre. Taking all three activities, eight boys undertook all, fourteen boys one or two, and seven boys none. The mothers of the latter explained: 'I have a terrible fear if he went on his own something would happen.' 'I wouldn't ever let him cross the road, you couldn't trust him - he'd get himself lost.' 'He'd get lost on his own, Robin takes him, twice he's been brought back by the police.' 'I don't like him going into the shops because of the temptation there - when they come from poor families the temptation is there; they only go if I go.' 'He didn't go when he was ten, he does now, not that I like him to - there's lots of temptation - I can't be behind him every five minutes.' *How do you feel about children going off to places alone - does it worry you at all? How do you think you can protect a child from being frightened or hurt by someone when he is on his own? Have you discussed this with him at all?* All but three parents expressed worry: 'With all that you see in the papers - children being taken off, and Margaret (the child-murder case), I like to see them home from school. When Margaret was missing the police came round checking and that did worry me.' Three mothers mentioned incidents in their own families: 'Ginette got taken away not so long ago. She was picked up in Heath St and was gone three hours, it was a big man in a big red car, he had a hat on. She was not interfered with.... It frightened her.' 'I won't let him go far on his own. Stephen, now, some man got him in the park, the police came and Stephen went to the police station.' 'A bloke followed Tina in the park, Tina told a woman who fetched her husband who told him to clear off.' Eight mothers referred to accidents: 'He fell in the river... and he was drenched, coat and all.' 'Yes, because of the railway - I warn and threaten them, but they still go back, every school holiday we get a letter reminding us of the lad who lost his legs.' Another mother referred to undesirable activities by other youngsters: 'Things are happening in the house next door (vacant prior to demolition) - it's terrible for children to see - boys and girls in there.' All these parents give their sons strong warnings not to talk to strangers, or to take sweets or money: 'I've shown them things in the papers when children get murdered.' 'If someone does talk to you go and see a policeman.' 'Ignore them, take no notice, tell them "here's me Mum".' Three parents however, took a different line: 'I have no fear for him, he wouldn't speak to anyone.' 'He knows his way ... he knows how to look after himself.' One mother worried about the younger children, not the 10 year old: 'I warn them not to speak to strangers, I can't see how you can protect them, you can't keep on protecting them, you have to let them go their own way, you can warn them, yes. Some people are not very nice, you have to look out, there's good and bad.'

Do you expect him to be in by any special time? Does he complain about that? All but four mothers have some rules about coming in. The latest time mentioned was nine to nine-thirty (three mothers). Six mothers insist that the 10 year old comes in before seven, the majority between seven and nine pm. 'Father makes him come in, there's no place to go after tea round here, you don't know what mischief they get up to at the park.' 'He's in at six - he doesn't like it but I know where they are, I don't like them running in the streets - it's a bad quarter.' Seventeen boys are said to accept these rulings, and another four have no reason to complain because there are none. Eight boys are more difficult and 'sometimes pull a job.' There is no consistent relationship between coming-in rules, degree of independence or the boys' willingness to comply.

When it comes to bedtime, however, the concerns that lead most parents into making rules no longer operate, the boys are home, they are safe. Asked about bedtime rules and complaints, the mothers' accounts fell into two groups: fifteen mothers said there were no complaints, fourteen had trouble now and then. The difference between the groups resolved itself on hearing that: 'He knows when he's tired and he goes to bed.' 'They go any old time now.' 'They are up early, they get tired, they go to bed early.' The complaints come from the families who expect the boys in bed at set times: 'I like Raymond to go up about nine to half nine - he gets there by half ten after all the arguments.' 'Yes we do have arguments, now I leave him up till nine or quarter past... he nudges me and says "can I slip down?".' 'Yes there are arguments if he wants to see TV, I usually give in.' 'He likes to stay up, he doesn't get enough sleep - he wants to look after me, I let him at present. If I put him to bed early he gets up and makes me a cup of coffee without being asked.' Only two mothers made quite clear that they enforce bedtime at a time they consider appropriate: 'If I tell them to go up, they go. I put them up together at about eight o'clock.' 'He says he doesn't want to go to bed, but he goes, he doesn't really complain, he knows I mean it.' Both boys belong to the seven who are not given much freedom of movement.

MODESTY TRAINING

Fifty-six children aged 3-4 years

This aspect of training brought out strong convictions among most of the mothers, in contrast to some of the other topics where there was more uncertainty or even indifference. Mothers were not unlike Nottingham working-class women in following practices which seem to be rooted in feelings which they were reluctant to express; and like Nottingham working-class women they tended to avoid explanations

130   Chapter 9

even if their children demanded them.

Does he play with his private parts at all? What would you do if he did? Table 9.5 shows that in the Nottingham enquiry middle-class mothers are much more reluctant to use punishment in this situation than the working classes.

TABLE 9.5 Reactions of mothers to genital play (percentages)

	Nottingham						Socially handicapped families
	I&II	IIIn-m	IIIm	IV	V	All	
Mothers							
ignored or permitted	17	11	10	5	10	10	16
discouraged, not punished	78	71	54	54	42	59	39
punished or threat of punishment	5	18	36	41	48	31	45

Nearly half the mothers in our study deal severely with this behaviour: 'It's a rude thing to do especially in front of little girls.' 'I'd stop her, I don't think it's nice when there are big lads about the home.' 'I say, you dirty devil.' But the proportion who ignore it indicated they thought the children were too young and 'innocently' engaging in this behaviour. 'They'll get out of it.' 'He packs it up when I take no notice.' 'He's too young to know.' *Of course most children go through a stage when they think anything to do with the toilet is terribly funny. How do you feel about children giggling together over that sort of thing? Would you discourage it or just take no notice?* About half the Nottingham working-class mothers and 60 per cent of professional-class mothers (I and II) ignore this behaviour; 57 per cent of mothers in our study also ignore it. Among the rest it does not call forth strong feelings, many said they did not have to face this kind of behaviour, but would stop it if they did. Only one mother would smack. *What about children wanting to go to the toilet together or wanting to look at each other when they are undressed?* The answers in the Nottingham study show a strong polarization of attitudes: class I and II mothers are very permissive with 62 per cent allowing sex interest, and only seventeen imposing restrictions, whereas only a quarter of class V mothers allowing and half imposing restrictions on sex interest. The proportion of mothers who impose restrictions is even greater in our sample. Seventy per cent indicated strong feelings although – or perhaps because – they live in conditions which invite easy mixing of the sexes as boys and girls frequently share a bedroom. 'The girls undress down here. They never dress or have a bath in front of one another, Father doesn't like it. He says they will know when they are older.' 'They don't take their knickers off, not in front of one another, I keep them separate when I bath them.' 'The girls are usually washed first and finished by the time the boys come, the boys wash themselves.' 'I don't allow that - Tony gets undressed at the back if Karin is down. I don't like them sleeping together at this age' (these two children, aged 7 and 10, have to share a bedroom). 'It's hard for Mary - she has to stand on the landing and get undressed, she doesn't like going in the room with the boys. The eldest boy is a tease.' 'The boys sleep in underpants.'

Of the fourteen mothers who do not impose special restrictions only four have teenage children. Many said briefly they did not mind if children of the opposite sex see each other undressed. 'They go in one bedroom, so you couldn't mind about them seeing one another undressed.' Only one mother seemed to be at ease: 'Our Joe (aged 7) has a giggle sometimes, we don't take any notice. I think if you take no notice they don't get conscious about it. It's more natural I think if they don't get funny ideas.' *Does he ever see you or his daddy undressed?* The percentage of Nottingham mothers who impose restrictions in this situation rises from 19 among middle-class families (who are likely to have private bedrooms and a bathroom) to 51 among skilled, 54 among semi-skilled, and 68 among non-skilled families. The percentage in our sample was 70. 'I'd feel embarrassed. 'I was brought up that way, my own parents have never done it.' 'Father doesn't like it, it's disgusting he says.' 'We would turn our backs on them.' Eight mothers allow their girls to be present when they are undressed, but not the boys. Among nine mothers who 'do not mind either child', some said: 'He doesn't take a bit of notice'; 'Nothing wrong with it, they're only babies yet, round about ten I wouldn't like it but it doesn't bother me now.' 'I think it's part of being a family - it should be natural.'

Does he know where babies come from? (i.e., from mummy's tummy?) Would you tell him if he wanted to know or do you think he is too young? What would you say if he asked you? The Newsons' objective in formulating these questions was

> to distinguish those mothers who are prepared to take what is normally the first step in explaining human reproduction to the child from those who prefer to be evasive or tell a direct lie; and to leave aside the question of whether these more informative mothers would be prepared to follow up the basic information with further details (ibid., p. 376).

We cannot assume that, if a child in our sample already knows at the age of four, it is an indication that their mothers have explained where babies come from. The children mix a great deal with neighbours' children of all ages, and all have older siblings. Many mothers were vague about their young children's knowledge because they could well have obtained information from others in the family. Table 9.6 showing 18 per cent as believing that the child knows, and 30 per cent as willing to tell him if he asked, differentiates between two not dissimilar groups of mothers. The more significant division is between these two groups on one hand and the 52 per cent of mothers who would not tell yet. This latter group is in line with Nottingham working classes whose reluctance to discuss these matters indicates strong sub-cultural taboos. However, on enquiring from mothers 'who would not tell the child yet' whether they avoid the child's questions without giving a false explanation, the mothers in our sample reacted very differently. While Nottingham working-class mothers, who do not like to discuss the matter, prefer to give a false explanation, the percentage of mothers in our sample who avoid talking about it without adding a false explanation is much higher. The reason may be the large family size; pregnancy is a common condition of which the 3-4 year olds are mostly quite aware: 'Last night when I was so bad with my back, he came up and put his arms round me. He worries when I'm not well - he asked me what was wrong. I told him it was a baby in my tummy and that it was kicking.' 'When Dawn was born Janet was in the room, she held my hand all the time. When the time came for Dawn to be born the midwife told Janet to go downstairs, but Janet said "piss off" and wouldn't go.' 'She knows, sometimes when she's had enough to eat she says "I have a baby in my belly".' But uncertainty is widespread: 'From what I've heard from what the children round here say, I think they all know.' The mothers who would not talk to their young children expressed strong feelings: 'I wouldn't let them know, I don't think it's nice. If they asked, I'd say "God brings a baby - you've got to be good". I would never tell them, I'm too backward.' 'I don't agree with that on the telly - learning them at seven or eight is too young - if he asked me I'd change the subject.' 'I wouldn't tell, I don't really know, I was never taught it, I just picked it up. This sort of thing is never discussed in Ireland, I'd find it difficult to talk about.' When mothers resort to false explanations, which 29 per cent do, the commonest version is that babies are bought or 'I'd say they come from hospital', or 'they are born in a head of cabbage or in a turnip, that's the way I was brought up'. 'They are fetched in a black bag.' Two mothers mentioned the stork.

TABLE 9.6 Attitudes towards telling child the basic facts of reproduction (percentages)

Class:	Nottingham						Socially handicapped families
	I&II	IIIn-m	IIIm	IV	V	All	
1 Child knows	44	26	15	14	15	20	18
2 Mother would tell	44	47	30	39	10	34	30
3 would not tell yet	12	27	55	47	75	46	52
Category 3:							
Mother would not tell yet							
a avoids question	4	8	14	7	9	11	23
b gives false explanation	8	19	41	40	66	35	29

The Newsons sum up their findings on modesty training by stating that there are sharply polarized attitudes and quite separate philo-

sophies of behaviour at opposite ends of the class scale:
> At the upper end of the scale parents subscribe to the view that it is natural, and therefore right and proper, for children to be curious about their bodies... and to want to know the sources of new life.... At the other end of the scale the contrasting philosophy is that sexual curiosity is suspect, sexual information dangerous and perhaps frightening for the young child, and both must be controlled by being suppressed (ibid., pp. 384-5).

This is not an inaccurate summing up as far as our sample is concerned, even though they are not entirely comparable with 'the other end of the scale' of the Nottingham sample. The majority of our mothers felt that things to do with sex are 'rude' or 'not nice' and should not be discussed. Avoidance of the subject was usually coupled with a hope that someone else will do the job of passing on the essential information. The Newsons believe that in one respect the classes agree, and that is in preserving childish sexual naivety, but that the method differs by which to achieve this objective: 'Professional-class mothers seek to neutralize sex interest by bringing it into the open... the unskilled manual-class mother seeks to suppress it by outlawing it from the outset' (ibid., pp. 385-6). There are many mothers who will take considerable trouble to avoid being questioned, but they do not take their endeavours to conceal sexual matters to the extreme of deceiving their children. The percentage of our mothers doing so is less than half of unskilled Nottingham mothers, and even less than the more skilled sector of the working classes. There are in our sample enough exceptions to examine the attitudes of those mothers who, like Nottingham professional-class mothers, 'seek to neutralize interest by bringing it into the open.'

The study of families in Boston (Sears, 1957) found that mothers' 'permissiveness' in one area of modesty training was fairly closely related to permissiveness in others. The intercorrelations ranged around 0.60. There is no comparable index in the Nottingham study. We designed an index of modesty training based on responses to five questions with a maximum score of eleven. The following pattern emerged:

Very restrictive	(0-3)	17	(30%)
Restrictive	(4-6)	30	(54%)
Permissive	(7-11)	9	(16%)

Very restrictive mothers scored almost uniformly low on all five questions; the permissive mothers were equally consistent. Minor variation occurred on 'giggling over the toilet', genital play, and not liking to be seen undressed by the children. Permissiveness in modesty training will be further discussed in chapter 11.

Twenty-eight boys aged 6-7

Mothers, on being asked whether their sons had any tiresome habits *such as playing with his private parts?*, did not seem to be preoccupied with this behaviour. Only two mothers commented: 'The friend he's with at the moment, I'm not keen on him - he talks ever so rude, and because his friend does it, he does it, mainly when he's been told off.' The second mother did not enlarge but told us that she tells him to stop. Both are in the group of restrictive mothers.

In answer to *What about the question where babies come from? Does he know about that yet?* Eleven mothers said their sons did know, one mother was not sure, and another four indicated that a 7 year old is too young to be told, but that they were prepared to tell them later. A group of twelve mothers, on the other hand, said their sons had never asked and they were not going to do anything about it: 'I haven't thought about it - I think they are taught in school. I'm bashful with the boys, I like to tell my girls when they are older.' 'They seem to find out at school, I don't think I could tell them, I leave it to Dad when they are about 10 or 11.' 'School programmes are a bit alarming to me, all right for the girls, yes, but not the boys at that age. It's most important for a young woman to learn - I don't think boys should be allowed to see an actual birth of a baby. I don't mind them learning other parts.' While this mother's attitude in talking to us was relaxed, other mothers had difficulties in discussing the subject: 'I'd be too embarrassed to tell them, I wouldn't tell them at all. Twinny is all for God - babies come from Heaven. John thinks it's from the second-hand shop.' 'I'm bashful with the boys.' 'I hope Father will tell him, not me.' The mothers who expressed these feelings were ranked as 'restrictive' or 'very restrictive' in modesty training of their pre-school children.

Twenty-nine boys aged 10-11

The topic was introduced by asking mothers about their sons' television viewing habits, and this led to the question: *When he has been watching TV does he ever ask you questions that you do not really want to answer? What sort of questions? Does he for instance know where babies come from?* (If he knows) *Did you tell him or did he find out from somebody else?* (If he does not know) *Would you tell him if he asked you, or do you think he is too young?* The introductory discussion revealed that fourteen mothers did not mind what the boys saw on television. Fifteen mothers felt uneasy. The objections they expressed were possibly suggested by the prompt in one of the earlier questions '... *such as violence or lovemaking*'. Seven mothers expressed a dislike of lovemaking programmes, three of violence, and another five of both: 'I don't believe in them sex programmes.' 'I don't mind violence but I don't let them watch lovemaking, I turn it over.' 'This thing they put on about sex and that - I like them in bed for that - he's not bothered about violence.' 'I get annoyed at the language they put on, I try to avoid horror films and lovemaking.' Not counting objections to violence, twelve mothers (41 per cent) expressed strong adverse criticism of sex programmes; six of the mothers were earlier ranked as 'very restrictive' and six as 'restrictive'. The mothers who do feel uneasy about unsuitable programmes are in a quandary about the kind of action they should take, as television is the family entertainment laid on and seen by everybody in the family living room. 'I don't like him watching violence, but if it's on, it's on, then you can't turn it off, can you?' Thus, of the objecting mothers only half manage to impose viewing limits, the others tend to ignore the matter.

Eighteen mothers (64 per cent) said their boys 'never ask any awkward questions'. This may be because the boys have learnt that awkward questions are not answered. Eleven mothers said that the boys sometimes ask awkward questions, but not all had sexual topics in mind: 'For instance, about spastic children, I wouldn't watch it, but he watched it. "Why do things happen like that?" he asked.' 'Paul saw "The Hunchback of Notre Dame" on the telly and he asked "Is it real?".' But eight mothers referred to sex: 'If he sees a man and woman together'; 'about romance and such like'; 'about birth of babies'. All but two use evasive techniques in answering: 'If I don't want to answer I tell him he's too young at present, these things are for grown-ups to know.' 'Certain questions I try to avoid, I try to put it in a roundabout way.' 'I walked out - I was embarrassed.'

Nineteen mothers said their sons knew about the birth of babies. Three had either witnessed births at home or had been told by their parents, the others from sources other than the family: 'When they brought "Sex for schools" out on the telly.' 'Now they have the school programme - at first I was frightened as to how he'd react, he came home dinner time very quiet, he said "I'll tell you about it tonight" - I said, "don't be ashamed, it's natural".' 'He must be hearing it in school, I didn't tell him, I wouldn't like to tell him, I would be embarrassed, I think it's the way you have been brought up.' Some mothers said the boy 'had picked it up' or 'found out'. Three mothers could not say if their sons knew. Seven mothers said they believed their sons did not know, six of them thought he was too young: 'I leave it to Father to tell him or they learn it at school when they are 13 or 14.' 'He's too young, I might tell him when he leaves school - I think it's filthy.' 'He's never asked anything like that, maybe he learns it at school, I'd prefer him to. It's better when he is among a lot of children rather than on his own.' Of the mothers who expressed embarrassment three had been classified as 'permissive' in handling the preschool child; this evidence strengthened our interpretation of permissive practices in modesty training. In many cases it reflects a relinquishing of parental responsibility in spite of strong feelings, and not a practice adopted because of strong convictions.

JOBS AROUND THE HOUSE

Twenty-eight boys aged 6-7

We would like to know about the sorts of jobs the children do around the house at this age. Is there any little job you expect him to do now? Is that something he does as a regular thing, or just when he feels like it? Suppose he is too busy doing something of his own one day - what happens? The results from the Nottingham study are given in Table 9.7 together with our own. The proportion of boys who have regular duties is much the same across the social classes in Nottingham, about one-quarter being expected to do certain jobs. It is somewhat lower in our study. But there is a social-class trend in the percentages of boys who are not expected to help at all, which increases as social status declines; our figure of 61 per cent may be seen as an extrapolation of this trend.

TABLE 9.7 Jobs around the house (percentages)

Class:	Nottingham						Socially handicapped families
	I&II	IIIn-m	IIIm	IV	V	All	
Regular duties	25	32	28	22	25	27	18
Some help	32	33	25	27	19	27	21
None expected	43	35	48	51	56	47	61

Only five mothers in our study said the 7 year old had regular duties: 'He has a turn at tidying the bedroom once a week, and also empties the bucket once a week. Sometimes he will help with the spuds. If he's busy I say to him he can find five minutes.' 'He makes tea and fetches the shopping regularly.' 'He'll clear and wipe the table for me, he sweeps up, he fetches the coal - I don't expect him to do a lot - every morning he does it, he puts cups out on the table, he's down before anyone - he'll go down for coal without being asked.' 'He makes his bed, tidies the bedroom, fetches coal in - that's done regularly by the boys. If he's busy I still make him, he can go back later on, or I tell him to do the job when he's finished his own.' 'I like him to help with the table, he likes to help me regularly. I don't mind when he's busy, I never ask him, he does it for himself. He likes having a scrub at the washing and holding the pegs.' The last quoted mother expressed a feeling which is widespread among the families: she likes the boy to help, but she will never ask him. The mothers who said the boy gives 'some help', six in all, expressed similar attitudes: 'He usually brings the wet sheets down in the morning - one of them will do it.' 'He fetches coal, he goes to Mason's - though I'd rather send Valerie, she's got a better head on her shoulders. He clears the table and clears the toys up regularly. If he's busy I leave him and do it myself.' 'I usually try to make them all do a bit, not one too much. He does the dishes, clears the table, sweeps occasionally. If he's busy I leave him.' 'He collects wood for me, he goes to the shop for Dad to get his paper, if he's busy, Jane goes.'

Is he the sort of child who usually agrees with what you want him to do, or does he tend to object to things quite a lot? Many mothers chose one of the suggested answers: thirteen boys were described as 'easy', three as 'variable', and twelve as 'tricky'. Some added further comment: 'Stephen is a loving child... he'll do anything for you, clean the table, sweep - he'll do these things. But at the moment he is going through a stage when he's telling lies.' 'John is easy to manage in his ways, but if I ask him to get undressed for bed he objects, any other time he's all right.' 'He's quiet.' 'He'll do anything for you.' 'He isn't bothered.' *What happens if he simply refuses to do something you want him to do?* The objective was to ascertain how persistent mothers were in making demands. The Nottingham study showed no significant class differences, but a greater perseverance in making demands backed up by physical punishment, than did our study (Table 9.8).

TABLE 9.8 Mothers' reaction to child's refusal to help (aged 6-7) (percentages)

Class:	Nottingham						Socially handicapped families
	I&II	IIIn-m	IIIm	IV	V	All	
Mother							
ignores or makes no demand	6	9	6	7	9	7	25
rebukes	45	38	34	39	37	37	50
smacks	49	53	60	54	54	56	25

A quarter of the mothers in our study make no demands on the boy or ignore his refusal, indicating that they do not expect a 6-7 year old boy to help: 'They never do anything, I wouldn't hit them a lot, they have enough when they get older. If he refused, I'd send someone else, but he's a good one, Sam is.' 'It doesn't really happen - but I'd do it myself.' 'If he doesn't I'd rather do it myself and don't ask him twice.' 'I just leave him.' 'They never do anything, they have enough to do when they get older, they are entitled not to do anything.' 'I wouldn't make Twinny do nothing, not any of my weens.' Mothers who make demands and insist on them being carried out, mentioned duties such as taking messages, shopping, washing up, clearing the table, wiping the table, clearing up a mess they had made or getting the coal. 'I raise my voice.' 'I'd give a little shout and I might have to keep on and say I'll slap him.' 'I ask him to get the coal in, then I shout at him.' 'Sometimes in the morning he's slow going to the shop and he gets a telling off. He's slow over washing - he's just playing - I tell him off or smack him, he's old enough to understand.' 'I have no real trouble with him, he gets annoyed with me if I keep on sending him to the shops when I've forgotten something - he says "Can't you get it yourself?".' 'I give him a good slap, if they get away with it once they think they can always get away with it.' 'I'd bash his head in - if he wouldn't - I'd kill him, they don't get away with nothing.' 'I'd lose my temper and hit him - I'd jump at him with a shoe.'

Does he ever help you by looking after a brother or sister or the younger children? Over half of Nottingham's class I and II families do not let the 7 year old boy look after younger children, and the percentage of working-class families is about 40. Ten mothers (36 per cent) in our sample do not: 'I wouldn't leave him in charge - he'd kill them, he has no patience.' 'No - I never force them to do anything.' The percentage who will leave the boy in charge for a short time is 43, and for longer, or away from home, is 21. Taking both categories together, this percentage is not very different from Nottingham working-class families. 'He's always helping with the younger ones, he could take charge if I asked him to. I leave him in charge in the house - he doesn't take them out, the road is too busy.' 'When he comes home from school, he'll watch the babies while I pop out.' 'He likes to sit and nurse the baby until she wets him, then that's that.   He takes them out to play with him.'

Twenty-nine boys aged 10-11

What jobs do you expect him to do for you now - I mean without being paid? Seventeen boys (59 per cent) help around the house: 'He keeps the backyard nice and tidy - the kids throw rubbish in. He washes up the tea cups.' 'He runs messages and helps with the children, he tidies the bedroom. I don't *ask* him to do the cleaning, but he will.' 'He brings the coal in, he washes up, he'll do messages. If I'm not well he'll make a cup of tea.' 'He's handy with little jobs, he fetches coal and sweeps the yard. He covered the sideboard with "contact".' 'He likes to make the beds, sweep the room and chop wood, he'll wash up - he likes to do it by himself, he doesn't like you there.' 'I don't pay him to do nothing, if I ask him he'll do it - mopping, cleaning - he'll do anything. He'll come and ask me if he's done a thing right, if not, he'll go and do it properly.' 'He does practically everything at the moment - I don't want to go to hospital - he will wash dishes, do the mopping, the dinner - that's the beans - he'll make coffee or fry eggs.' In all these statements the boy's willingness to help, or even liking to help, is expressed. Seven mothers said their sons would do certain jobs, but not regularly, and they are not made to do any: 'He'd put the kettle on now and again, and he'd run over to the shop.' 'I wouldn't ask him to do the housework, but he'd go errands, fetch the paper.' Some of these boys make their own beds, and they are expected to keep themselves tidy. But five mothers do not expect their boys to help at all: 'I don't expect him to do too much, he'll tidy up - he won't put his clothes away when he gets undressed.' 'I want him to fetch messages but he won't.' 'John (younger brother) will wash up, but Raymond won't, he doesn't do much, but he loves gardening.'

The pattern of expectations is similar to that observed in that half of the sample of families who have a 6-7 year old boy; many jobs need doing and the children share them out. Mothers do not allot particular jobs to individual children according to age and degree of responsibility, but in a number of families certain jobs are always done by specific children. Almost two-thirds of the boys aged 10-11 are helpful around the house; we have no comparable data from other social groups. The view traditionally attributed to working-class families that work around the house is woman's work is not borne out by these facts. We tested the willingness of boys to help against absence or presence of older sisters.

	Older sister(s)	None	Total
Regular jobs	9	8	17
Occasional	5	2	7
Never	5	-	5

Of boys with older sisters only half will help, whereas of those without three-quarters will do so, but the trend cannot be shown to be statistically significant.

A further question tells us more about the voluntary nature of the boys' jobs. *Suppose you ask him to do a little job for you, does he take a lot of trouble to finish it carefully, or do you think he would let it slide if he could?* All but two of the seventeen boys who help regularly are said to be reliable. 'I don't ask him to do the cleaning but he will do it provided he's on his own - if he is with anyone then he gets very slow.' 'He runs errands, he minds the younger ones, he makes me a cup of tea - he usually does these things if I haven't asked him.' 'He's pretty good, if I tell him I don't feel so good, then he does it - he cleans the yard, shops, he makes tea, he puts coal on the fire.' The seven boys who do not help much are described in different terms: 'If he does a job, he'll do it all right, but he rarely does a job - if I have money, I pay him, I think it encourages them.' 'He wouldn't do anything unless I kept at him.' The theme of helping voluntarily is taken up in a question asked earlier when mothers discussed their relationship with the 10-11 year old. *Is there any special thing about him that gives you a lot of pleasure?* Eleven mothers spontaneously mentioned their sons' helpfulness: 'He's a good kid - he's helpful.' 'He's helpful - we get on well together.' 'He's very helpful, he's good, he doesn't moan if I ask him to do things. When I come in he makes me a cup of tea and then he asks me what I want him to do, he knows there's a lot to do.' 'He likes to be in the house with me, he helps doing the housework.' 'He'll do more for me than the girls will - I get on better with the lads, I can leave him to do a job - he'll do it.' 'He helps with the babies, he buys me chocolate if he earns any money, he's good-hearted.' When mothers were asked *What do you think is his biggest fault?* not one mentioned disobedience on being asked to do a job. Most mothers complained about temper, arguing or fighting with siblings, impatience, or 'fidgeting'.

SOCIAL LEARNING AND CONTROL OF AGGRESSION

Fifty-six 3-4 year olds

Almost all children play with other children, either siblings or friends at least some times. The percentage of children in the Nottingham study who never play with others is seven; in our study this is an unknown situation. Company is the training ground for social behaviour, but play with others does not guarantee that the child adjusts his behaviour automatically to fit in with that of others. The sharing of possessions, the fair allocation of a much-desired object, patient waiting for one's turn, are behaviour patterns that need to be learnt. The Newsons found that many Nottingham mothers see playing out as 'providing a useful foretaste of the hierarchies of a competitive society' although not necessarily providing a civilizing influence on the children. Many mothers in the professional classes emphasize the importance of 'keeping a close watch on their children's play' and their choice of friends and 'peaceful coexistence' is seen as the main learning objective rather than 'survival through aggressive competition' (ibid., pp. 108-9).

Mothers were asked: *What do you do if there is a disagreement or quarrel? Do you think children should be left to settle their own differences at this age or would you interfere? Supposing he*

*comes running to you complaining of another child what do you do?
Do you ever tell him to hit another child back?* Maternal responses were classified in three categories:

1 mother accepts role as arbitrator, she interferes in order to ensure that justice prevails, she is prepared to hear both sides of any dispute and attempts to discover the rights and wrongs of the case;

2 mother's attitude is balanced between 1 and 3, she may prefer one attitude but finds it impracticable, or there is insufficient information to categorize;

3 mother prefers to let the children settle their own differences, she believes that children must learn to fight their own battles, and for this reason she does not usually listen sympathetically to complaints and tells her children not to come telling tales.

As shown in Table 9.9 the undecided group of mothers formed nearly half the Nottingham sample, but there was a significant social class difference of attitudes in the two groups of arbitrators and non-intervening mothers, and this is strengthened by the findings of our study. The undecided group is of the same magnitude as all Nottingham mothers. The group who will arbitrate is most strongly represented among class I and II Nottingham families and reaches about

TABLE 9.9 Mothers' attitudes in children's quarrels (3-4 year olds) (percentages)

Class:	Nottingham						Socially handicapped families
	I&II	IIIn-m	IIIm	IV	V	All	
Mothers arbitrate	37	25	21	18	20	23	(4)*
half/half	44	50	46	44	37	45	44
children settle differences	19	25	33	38	43	32	52

* The number was only two.

one-fifth among working-class families. Only two mothers in our sample said what may amount to 'interference in order to ensure that justice prevails': 'I would see what's causing the argument, I'd go out and see what they had done.' 'I could give out to my children, telling them off - I couldn't give out to other children.' The largest group was that of mothers who have no doubt about letting the children settle their own differences; with 52 per cent it was well above the percentage of Nottingham working-class mothers of that opinion, and it obviously reflected the feelings of women in densely populated inner-city areas. The reasons for their feelings are training for independence, a reluctance to listen to frequent complaints, and a desire to avoid quarrels with neighbours over their children's behaviour. 'I let them sort it out for themselves - it makes them more independent.' 'If you interfere they are only at it again soon, he has often come in complaining about children hitting him, and I've gone out and there's been no one there.' 'They should learn to stick up for themselves.' 'If you fight their battles they are going to come to you all the more, even when they are grown up.' 'If you stick up for them, they won't stick up for themselves.' 'All you do is to get the other mothers on to you, you have to let the kids fight it out. Of course when there's a bad cut or something the mothers come with a mouthful of abuse to you - I've had it happen - "your child attacked mine first" and all that. You have to work it out yourself without causing arguments - I'm a peaceable person.' A number of mothers would intervene 'if they went too far, if they started knocking each other to pieces'; 'if blood runs'; or 'if they use bricks and stones'.

The same social class difference emerged in Nottingham in comparing mothers' encouragement of self-defence. As shown in Table 9.10 mothers who never tell their children to hit back dwindle from over a third in the professional class to four mothers in our study. The Newsons point out that middle-class mothers have the advantage to withdraw their children from neighbours or to send the neighbours off and close the garden gate. The working-class child in the inner city has to encounter the 'jungle of might-is-right' and self-defence is a necessary method of existence.

TABLE 9.10 'Do you tell your child to hit back?' (percentages)

Class:	Nottingham						Socially handicapped families
	I&II	IIIn-m	IIm	IV	V	All	
Never	35	16	16	10	17	18	7
yes, qualified	29	20	18	28	18	21	41
yes, unambiguous	36	64	66	62	65	61	52

Nevertheless, mothers in our study more frequently qualify their advice to hit back than either middle-class or working-class Nottingham mothers, again probably in the awareness of close proximity of neighbours, frequent quarrels and their inability to send undesirable playmates home. The 'qualified' advice to defend himself is the nearest these mothers get to intervening in their children's squabbles: 'I go out and see what it's all about. If the other child were one to bully, I'd tell her to hit back, I don't like bullying.' 'If they are the same age I don't bother, they've got to learn.' 'I tell mine to hit back if they are in the right, you have to round here.' 'I do tell her to hit back if she's been hit, the big kids often hit the little ones. It's more difficult with neighbours' children - you can't get on to neighbours' children or be hard on them, it would insult their parents.'

Over half the mothers encourage the 3-4 year old to hit back without reservations: 'If someone went up and hit him, I'd tell him to hit them back.' 'If a child hit Dawn in the yard, I tell her to hit back - he bit Dawn, so I told her to bite him back.' 'I tell them to hit back, I think this learns them. Tina and Charles never take their own part - I had Chris with a broken jaw and nose because he wouldn't take his own part. I'm trying to teach the little ones something different.' 'You could stand on the steps and kids would thump *you* - they hit for no reason at all. It's hopeless fighting about kids round here, I used to have quarrels with parents over their kids, but now I leave them to it.' 'If people come complaining to me I just shut the door and don't listen.'

The number of mothers who told us that they do not hesitate to get involved with other mothers is only six (11 per cent). 'If it goes a bit too far I'd interfere, otherwise I let them fight it out themselves. I'd go and see what was happening. I'd go to the child's mother. We get a lot of arguments over the children - I had rows and arguments over kids these past nine years.' 'If you go to a neighbour they won't have it that their child has done any wrong, it's always *your* child - there have been six or seven arguments over children in the past fortnight.' 'There's a kid in the yard that keeps hitting Ivan, so I've told his mother I'll hit him if he does it again.' 'If there's a quarrel outside the family I'd fetch him in, I'd rather avoid trouble, but some time ago the lad over the wall hammered him on the head - he did it for nothing - I did go round to his mother for that.'

Twenty-eight boys aged 6-7

Do you ever interfere in his quarrels and arguments with other children outside the family? And do you ever tell him what he should do in his quarrels or help him to manage them in any way? In the Nottingham sample a quarter of professional-class mothers and over half of the mothers from unskilled families would interfere personally. This information must be interpreted in the knowledge of differences in environment; the friends of class I and II children are likely to be invited into house or garden, the friends of class V children play with them in the streets and yards. The percentage of mothers in our sample is 54, which is the same as among unskilled families in Nottingham. Their intention is the same as expressed in relation to the pre-school child: to prevent serious injury. 'If he was too rough.' 'If a bigger one picks on him.' 'If he is kicked by the one with the boots.' 'The young fellow next door hit him with the sweeping brush - he's older than any of them.'

When it comes to sorting out their sons' squabbles by counselling justice the proportion of mothers in Nottingham is rather lower: about one-fifth of middle-class mothers and only 8 per cent of working-class mothers said they did. None of the mothers in our sample said they did. The percentage counselling retaliation is about one-quarter right through Nottingham social classes, and it is 54 per cent in our sample: 'His father tells him to fight back, but not to pick bricks up, just use fists.' 'I tell him to hit back, I don't like kiddies being timid.' 'I try to learn him to defend himself.' 'Father tells him to hit back, he says "If you can't use your fists, use your feet - use your nut".' The percentage of Nottingham mothers who counsel withdrawal or diplomacy is thirteen with no class differences; only two mothers in our sample did so: 'Sometimes I coax him not to fight - I say "Charlie, don't get me in trouble, just be friendly".' About one-fifth of Nottingham mothers, and about one-third of mothers in our sample refuse to be involved in their children's quarrels.

Chapter 9

Twenty-nine boys aged 10-11

Outside the home the boys of this age group are less easy to supervise than the younger siblings; the questions eliciting mothers' involvement are therefore confined to quarrels in the home. Judging by answers to the question *How does he get on with his brothers and sisters?* there appear to be daily quarrels among most families which mothers seem to take as natural; 'they are always on to one another', 'he tries to boss the babies', 'she keeps teasing him', 'they're like cats and dogs'. Only seven mothers said there was not much quarrelling; the children in these families appeared to us to be quiet, shy and somewhat withdrawn. The kind of situation that is typical is described by mothers on being asked *What makes them really angry?* The answers indicate the irritations experienced when siblings fight over valued possessions, or when they have to share very limited space for play; but other incidents are of a more universal nature. 'He wants a ball and Bernard won't give it him.' 'When the others get something and he doesn't.' 'If Donald borrowed his football shirt – up in the air he goes.' 'If the kids have one of his cars, he's always angry.' 'If he's done something and Tommy or Joe disturb him he'll go to smack them, but then pulls back and starts shouting "I've been doing that for nothing".' 'Anybody like his brother tormenting him.' 'Interruptions on the telly.' 'It's the little things that bother him – he's irritable, little things like others winning a game.' 'Children calling him names.'

Ten mothers reported situations which reflected shortcoming of themselves, or criticisms made by the boys about their parents: 'If I don't listen to Terence.' 'If he doesn't get his money.' 'If he has to wait for things like tea or soup.' 'When I shout at him.' 'When he has to go to bed very early.' 'If the others don't help with jobs – if he's been making tea during the day and I ask him again in the evening and there are others around who could do it.' Or, most self-incriminating: 'If I pick on him for something he hasn't done.'

When he is upset does he take it out on other people or on himself? Does he ever blame other people for things that are really his fault? Just over half the mothers, sixteen in number, thought the boys tend to take it out on others, often going for the younger children: 'He'll hit the little ones if I've upset him.' Two boys are described as getting into a temper; 'he pulls things off the settee', and 'doors bang and I've even had him throw chairs over'. Eleven mothers, on the other hand, described the boys as tending to sulk: 'he won't speak', 'he'll sit and cry'. Four of these eleven will never blame others when they are at fault: 'He's very honest – if he's done a thing he'll say he has.' 'He'll admit it's his fault.' 'He's truthful.' 'He'll take the blame'. The rest will blame others at least sometimes, if not frequently.

When there are daily arguments, quarrels and irritations from which nobody can escape except by leaving the house, the role of parents as arbitrators becomes that much more important. Mothers had an opportunity to describe their efforts to disentangle wrong-doing and to punish the wrong-doer when asked *When he is in a rage (shouting or sulking) how do you get him back to normal?* Only one mother said: 'I find out the trouble, I see what's going on.' Nineteen (34 per cent) take no notice: 'I ignore him', 'I just leave him, he soon forgets', 'I let him work it off himself'. Five mothers make efforts to restore peace and quiet, but do not pursue the wrong-doer either: 'One of the younger ones upset him - he borrowed his wickets for a game of cricket, and his brother hid them and that upset him; or anybody like his brother tormenting him; or if he's touching something and it gets broken, he gets in a right tizzy. I tell him to keep quiet, and I try to find what he's looking for or mend what he's broken, or get another one if possible.' 'If Guy has anything of his that Dane really wants to keep he gets mad; they don't get a lot of toys, he likes to look after what he gets, he shouts and bangs - I soft-soap him, promise him this and that.' 'Garry gets angry when he can't have his own way - in the finish I give him his own way.' 'If Paul is hit he goes to town - I just talk and laugh at him and he has to laugh back.' Other parents punish indiscriminately: 'When anything doesn't go right for him he shouts and bangs and Father clouts him.' 'Sometimes he gets like that, I don't know what makes him like it, I shove him out the bloody door and he gets himself back to normal.' It appears that hardly ever is an attempt made to make the child or children understand the issues involved in a conflict and to help them sort these issues out. There is a tendency by mothers and fathers to punish all involved in a quarrel because of the noise nuisance to the family and neighbours, or to bring noisy scenes to an end by 'giving him his own way' or by 'soft-soaping him'. The circumstances are hardly inducive to disentangle squabbles and identify the wrong-doer.

SUMMARY

The children's experience of independence is a paradoxical combination of independence from their mothers at an age when children are considered too young by most Nottingham families, and restricted movement in situations considered normal in other families. Only about one-third of the 3-4 year olds are able to perform little tasks which about two-thirds of children of that age achieve among Nottingham families. The range of independent activities of the 6-7 year old boys is narrower, and the variety of activities of all age groups is restricted. In the home the needs of the family take priority over developmental needs of individual children. Injury must be prevented but it matters little who started the squabble, noise levels must be kept low, and consequently, if the parent feels it necessary to intervene, everybody may be punished. What little there is, is shared out; what needs doing is also shared out. Helpfulness is praised, but there is considerable tolerance of children who are not spontaneously helpful. While families varied in their approach to training in social responsibility, there appear to be clear divisions between families who insist on certain standards of behaviour and others who have a high degree of tolerance.

In some families all children play a part in helping around the house regularly, in other families the arrangement is more haphazard and the boys are not expected to do much except odd jobs, and there are some families in which the boys will not help at all, and mothers do not try to encourage them to do so. The numbers involved in each of these three groups were twenty-two families where boys help regularly, twelve where they will only do odd jobs now and then, and twenty-two where they will not help at all. While most mothers demand very little and many do not insist on obedience, there are some who do. We will discuss this aspect of parenting in chapter 11.

On an index of modesty training again the patterns that emerged differentiated a variety of regimes. Nearly one-third of the families were very restrictive, another larger group (54 per cent) moderately so, and a smaller group was permissive in the sense that they sometimes did not take measures to curb irritating behaviour, and at other times seemed to be indifferent to the problem.

The parental measure on which sharp contrasts in regime became noticeable was concerned with the degree of protection given to their children to ensure their safety. An index of 'chaperonage', designed by the Newsons, was modified for use in our study. It aims at measuring the degree of independent movement achieved by both age groups of school boys and the degree of contact kept between mother and child. The following items were scored for boys in the younger age group: mother does not fetch the boy from school, but he comes home alone (score 1); or he goes elsewhere (score 2); he undertakes three to four activities on his own (score 1); he roams around in the streets (score 1); mother cannot always find him (score 1); or often cannot find him (score 2); police record found wandering (score 3). The following items were scored for boys in the older age group: mainly outdoors (score 1); one or two activities on his own (score 1); or three or more activities (score 2); coming in after eight pm (score 1); or no rules about coming in at night (score 2); police record found wandering (score 3). On the basis of total scores the families were divided into three groups with the following distribution:

Much chaperonage	16 families (28%)
Some chaperonage	25 families (45%)
No chaperonage	15 families (27%)

'Chaperonage', it must be clearly understood, is a composite index which tells us something about the methods practised by parents to keep their boys under control, and about the degree of individual freedom of movement achieved by the boys. Obviously the boys who receive much chaperonage have little freedom of movement. This may prove to be important in an environment which, judging by parental accounts, teems with danger: we have given an account of mothers' fears about accidents in the roads or in vacated houses which are due for demolition. Many are worried about strangers who molest children prowling in the neighbourhood. Some mothers are worried that the boys may get lost, as in fact some boys have done. Some mothers find hostile neighbours a reason for keeping their children in the house or, if they have it, the back yard. Some are acutely aware of delinquent behaviour of neighbours' children and they do not wish their own to acquire it. Some mothers do not want their children to be involved in sex play at too early an age. And there are some who are depressed and lonely and who are comforted by the presence of their children. Between them, these forty-one mothers, in many cases with the support of their husbands, manage effectively to exercise some or even a considerable degree of control over their sons' movements. The fifteen families who do not chaperone their sons are by no means unaware of environmental hazards, and a number were in fact trying to keep their boys under closer control, but did not succeed. Others, however, trusted their sons' good sense, some confined themselves to warnings, two families sought the help of social workers when the boys were deeply involved in delinquent activities. We will discuss the consequences of this parental measure in chapter 11.

Chapter 10

TOLERANCE AND DISCIPLINE

The Newsons found that among Nottingham families strict training in obedience of the 4 year olds is no longer seen as an acceptable way of bringing up children; instead parents prefer 'voluntary and rational co-operation as opposed to automatic and servile docility'. But 'those who really believe that they never resort to authority are simply extremely subtle not only in disguising it to their children but in deluding themselves' (1968, p. 406). In spite of the change in general climate, three-quarters of Nottingham families were found to use smacking as an enforcement of compliance with parental wishes. Overall class differences are significant in those families that do not smack often (i.e., less than once a day): the wives of non-skilled workers smacked more frequently than the wives of professional men. Mothers who smack more than once a day, and, at the other extreme, mothers who do not like smacking at all, were not class-related. In a sub-sample of working-class mothers living on estates and others living in more overcrowded central areas of the city frequent punishment was associated with overcrowding. Class I and II mothers were found to be more consistent in the use of smacking; unskilled workers' wives, although in general more inclined to smack, showed less consistency in applying the punishment in particular situations.

SMACKING

Fifty-six children aged 3-4 years

What about disagreements? What sorts of things make you get on each other's nerves? Children of this age have a tendency to persist with certain forms of behaviour which they know irritate their mothers. Very typical among our sample - as in Nottingham - was whining and babyish behaviour: 'When she whines for something, and I haven't got it to give her.' 'When he's crying like a baby for nothing, I can't stand kids whining around me, especially if I know he's not hurt.' Several mothers complained about the mess young children make: 'When he runs in and out and messes the place up and throws stones and they fight with each other.' 'He plays with the water, and I could strangle him when he goes up and down the stairs and there's no call for it.' Other children pester their mothers to get something: 'If he keeps on demanding money and I haven't got it - he keeps on, even if I threaten him with the stick, he keeps on - I tell him "tomorrow", he says "no! - now!".' 'If you tell him he's not having something and he keeps on and on, he does a real war dance - "I'll tell my Dad" - he doesn't understand when I have no money to buy him an icecream.' Disobedience: 'If I ask him to do something and he starts dancing and kicking - if he gets in one of them moods.' 'If he won't do as he's told, I can stand so much and then I get nasty.' Undesirable habits: 'Diane sticks her tongue out and they both keep rocking.' 'She plays about with her food, especially on a Sunday, she messes with the potatoes and gravy, she doesn't mash them, she just messes.' This group of young children consists of twenty-nine girls and twenty-seven boys. The mothers of ten girls and of five boys said that the child did not get on their nerves: 'Inoffensive', 'not a lot to complain of', 'not a demanding child', 'usually you don't know he's there'. One mother said 'There's nothing really - I think I get on *his* nerves.' She had noticed that the question was two-sided.

A scale of frequency of smacking was developed in the Nottingham study dividing families into four groups:

1 never;
2 less than once a week;
3 once a week to once a day;
4 more than once a day.

We collapsed the four into two groups, as grading was found to be difficult. All families in our study agreed that they smack, but there was a distinct difference between those who smacked rarely and those who smacked frequently; three-quarters of the families belonged to the latter group, and in that respect resembled Nottingham working-class mothers.

Do you think it is necessary to smack children? The mothers who smack frequently (75 per cent) said that it is 'or else they get the upper hand'. 'You have to use your judgment, not smack them too often, but when they are really defiant you have to - if you don't, they walk over you as they grow older.' 'I've tried talking but it doesn't seem to help, the young ones see the older ones and think they can do the same and get away with it. It is difficult with a large family, it's hard to train them to be good.' That quarter of the group who rarely smack used different techniques: 'I try promising them something, I don't like to smack, not a lot. If he really does wrong I smack him, but to be too keen on it is bad, talking to him is better. You can either coax him or find out what he's done and why.' 'At Doll's age I don't think they know what you are smacking for.' 'I found that when my husband smacked them they would do more for me than for him.'

If he refuses to do something he really must do, what happens then? Almost half the group considered the 3-4 year olds too young to need much correction: 'He's too young, I leave him, I don't keep on at him, I tell him there's more in the family that will help.' 'He usually does what I ask him though I don't ask him too much.' 'If he's in that mood there's not much you can do about it.' 'I do it myself if I ask her to do something and she won't.' 'I let them carry on, I stand a lot off them.' Just over half, (54 per cent) said they would smack. A number of further questions referred to specific situations: when the child lies, when he tries to smack mother, when he has temper tantrums, and for genital play. As shown in Table 10.1, in some situations class differences become significant and are confirmed by our study. But mothers in our study tend to be more lenient when the child does not obey than working-class mothers in Nottingham, and this is in line with their attitudes towards the older boys, as described in the previous chapter. The relatively high rate of smacking for tempers, in contrast, may indicate the importance of keeping noise levels down in overcrowded conditions. *What sorts of things seem to start if off?* 'He has them whenever anyone upsets him, most nights, usually him and Karen. I have to get up and smack him, the quiet ones are always the worst in a temper.' 'Nearly every night, she stands and stamps her feet if she can't have what she wants when she wants it.'

TABLE 10.1 Use of smacking in specific situations (percentages)

Class:	Nottingham						Socially handicapped families
	I&II	IIIn-m	IIIm	IV	V	All	
Obedience enforced by smack or threat of smack	56	63	70	63	79	67	54
Child smacked:							
+ for lies	27	37	38	44	42	38	46
- for trying to smack mother	43	50	63	60	64	58	55
- for temper tantrums	20*	33*	36*	38*	25*	33*	51*
- for genital play	5	18	36	41	48	31	45

* Actual cases only.

Do you think smacking does him any good? Over half of the families feel it is an effective punishment: 'It makes them behave'; 'they know you mean it'; 'it learns them more, you waste your time talking to them'. The others were divided: fourteen mothers thought it was not effective: 'it doesn't get nowhere'; 'it doesn't affect her, she pays no attention'; 'they stop for half an hour, then they are back at it'. Another twelve mothers were quite emphatic that it is wrong; they smack rarely and only when they lose control: 'you knock one devil out and a dozen in'; 'it turns a child to hate you'; 'I was smacked when I was young, and it didn't get anyone anywhere'.

Suppose you asked him to do something for you and he said 'No, I can't do it now, I'm busy', what would you do? This question in the words of the Newsons, tests 'the mothers recognition of a principle

of reciprocity in her dealings with the child: her willingness to
allow the issue of fairness and democratic rights to prevail over
her wish to maintain authority' (1968, p. 394). This question
forms part of the Newson's subsequent interpretation of 'child-
centredness'. Answers were arranged in four categories, of which
the fourth, insistence on immediate obedience, is the least 'child-
centred', because the maintenance of authority is more important
to the mother than the child's play. The percentage of mothers
giving this answer in a Nottingham random sample was about the same
as that of our mothers, as can be seen in Table 10.2.

TABLE 10.2 Mothers' responses to child's excuse 'I'm busy'
(percentages)

	Nottingham random sample	Socially handicapped families
A - mother accepts excuse as valid	30	47
B - allows time but finally insists	26	14
C - accepts, but shows disapproval	13	5
D - insists on immediate obedience	31	34

A significantly higher proportion of mothers in our study, nearly
half, accept the child's excuse as valid. The comments made reveal
a very different situation and a very different approach: 'They
haven't got a lot of toys, so they don't really get pre-occupied.'
'I wouldn't bother, I don't force them, I've never made them do
things, I don't like arguing.' 'I don't force my kids to do nothing,
they tell me in this yard I'm too soft.' Absorption in play is not
a normal situation for many of these children, and obedience is not
demanded anyway. Some mothers, on the other hand, saw the situation
in the 'child-centred' way: 'It would depend on what they were doing
and how urgently I need it'; 'he has said that, he says it in a
comical way, I have to laugh. I'd probably leave him'; 'if he was
doing something I thought was more important than what I wanted him
to do, I'd leave him. If I thought he just didn't want to do it,
I'd make him.' Some allow time: 'I wanted him to go and buy a
packet of tea, and he wanted to play. I said, okay take your time,
but no circus please. It takes him about five minutes to make up
his mind, he says, "why always me?" but then he goes.' 'I treat
him more as a grown-up than as an infant, I tell him to hurry up,
I let him finish and then get him to do it.' A number of mothers
pointed out the relative attraction to the children of being sent
to the shops instead of hanging around at home: 'she'd be glad to
do it, if the others want to go, she screams'. The discussion on
child-centred behaviour will be taken up in chapter 11.

Twenty-eight boys aged 6-7 years

What sort of things make you get on each other's nerves? One-fifth
of the group said that nothing really irritated them. The others
described many kinds of naughty behaviour. 'He won't do as he's

told' was a common one; the boys insist 'and won't take no for an answer' another common one. They cry, nag, sulk, moan, complain. Some have tiresome habits: 'The way he eats – he'll eat all day long, he's always on at me for pennies.' 'Playing up – he throws knives and spoons.' 'He keeps running away all the time.' 'He sings around the bushes in the yard, the old lady next door doesn't like it, but he keeps on.' Picking quarrels and fighting over things with siblings was frequently mentioned.

How do you feel about smacking children of this age? In contrast with Nottingham mothers, of whom 73 per cent confessed they felt guilty or upset after smacking, only five mothers in our study indicated feelings of distress: 'I have to, I don't like it, it hurts me too... I could sit and cry afterwards.' 'I don't like hitting them, they have enough from their father, I feel terrible.' Three mothers said they never smack the 6 year old. The others are unemotional: 'With Stephen it's the only thing that has any effect.' 'I don't feel anything – none is too young or too old, they all get it alike, even the baby.' 'I can't say he's bad, but he just gets on my nerves, his ways irritate me.'

What sort of results do you think it gets with them? Only one-third of the mothers maintained it was effective; the majority admitted that smacking had undesired results: 'It makes him wild'; 'he gets cheekier'; 'he gets defiant'; 'I've smacked him a bloody lot, it don't get no results.'

Frequent smacking for behaviour which irritates mothers goes together with infrequent smacking, compared with Nottingham working-class mothers, for refusal to help, as reported in the previous chapter. Noisy and quarrelsome behaviour is found to be intolerable in overcrowded conditions.

Twenty-nine boys aged 10-11

Mothers were less critical of boys at this age than of pre-school children and the 6-7 year old boys. Nearly a third said they had no disagreements. Others commented: 'Only if he wants different food to what I give him – I say, if you don't eat that, you get nothing.' 'If I give one of them some money for an errand, and then he asks for money too.' 'His eating and eating – I have no money, and he'll eat a loaf of bread and he doesn't care if everyone else goes without.' Two boys wander off and come in late at night. Worry about delinquent behaviour was voiced by five mothers: 'When he stole the money.' 'When they have gone in the old houses.' 'If he goes to a shop and wants things he can't have, he just takes them.' Mothers smack the boys for persistent 'defiance', cheekiness, answering back, and kicking or hitting younger children in the family.

'I hit him all the time.' 'When I really get going I give him a hiding.' Only two mothers said they never smack the 10-11 year old, but one of them added 'I throw things'. Eight mothers smack very rarely, eleven more rarely, and eight very frequently. Fathers play very different roles; in nine families they leave the smacking to their wives, the others also smack with varying degrees of frequency. The majority of mothers said they have to be 'really mad' or 'really riled' or 'have to get balmy proper' to smack the 10-11 year old. Seven mothers use a belt or a stick: 'I got fed up hitting them with my hands, they take more notice of a stick, I keep it handy.'

THE USE OF LANGUAGE IN DIRECTING BEHAVIOUR

Fifty-six children aged 3-4

The Newsons found that one of the most salient ways in which patterns of upbringing differed between the middle and the working classes was in the preference given by middle-class mothers to the use of reasoning in the control of their children. Most families in our sample conformed to the patterns observed among Nottingham working-class families. There is a general disinclination to reason with a child in situations which need quick action; when the entire family has to share one room for daily activities, irritability runs high. But there are specific situations which mothers can handle in a calmer mood.

Suppose he says he hasn't done something naughty when you know quite well he has. What do you do then? Ten per cent of Nottingham mothers ignore this situation, and so do 10 per cent of our mothers. As shown in Table 10.1 (p. 149) nearly half the mothers smack the young child for lying, a rather larger group then the equivalent in Nottingham. But a group similar in size attempts to ensure that the child recognizes he is doing wrong: 'The other day he broke a cup and he placed the blame on Daren. I said, "I've seen you break it, pick it up" - he didn't want to. "I give you two minutes" I said, eventually he picked it up but kept insisting it was Daren did it, but then he did admit it. I told him he need not be frightened to tell the truth.' 'I make her tell me the truth, I tell her it's a sin and God wouldn't be fond of her - I don't believe in them telling lies.' 'I try to find out and make him tell me the truth - you can tell with him - he hangs his head.' In these situations the use of language becomes vital. In answer to a further question about cheekiness, only 5 per cent of Nottingham mothers were prepared to disregard this, in contrast with sixteen in our sample; about one-third had mixed feelings and would allow 'funny' cheek but not 'rude' words. Forty-six per cent smack cheekiness when it is taken too far, in contrast to 20 per cent of Nottingham mothers (there were no class differences).

Do you ever say he can't have something he likes - sweets or TV or something like that? Do you ever send him to bed? 'My husband is a bit firmer in that way, "no TV - bed", and he means it, I'll give him his due there. Men need peace and quiet after work; when we're speaking, he's firm with the children, but when we're not speaking he ignores them.' 'If she wants anything and I haven't smacked her and she's been naughty, then if the others have something I tell her why she can't have it, but I don't stop the telly.' 'I did that this morning, he tore a car about until the wheels broke off, so I had three toffees, one each for the other two and one for me. He screamed but then he said "I'll be good".' 'I talk to her, she is a fairly sensible child, she doesn't seem like a child to me, not since she started walking - you can talk to her like an adult. The other day she wouldn't keep still when we were out. I had promised her shoes but I told her, "No, you defied me". It's the only time I've backed down on a promise, but she'll get them at the weekend, so I haven't really broken it.' Some mothers are doubtful that it is an effective measure: 'If she doesn't behave I tell her she can't have sweets and I hide them, but she gets them in the end because she finds them. I have once sent her to bed, she kept jumping on the bed, so I brought her back down.' 'I say it, but I always give way with Rhona, it's wrong I know, but she's not a bad girl.' 'I send her to bed, she comes down five minutes later.' 'They always get their own way with the telly, I say it but I don't keep to it - Father says I give in too much.' But twenty-five mothers (45 per cent) considered these sanctions unrealistic: 'If you send him to bed he wouldn't go, he'd tell you to go first', or 'I've threatened to put them down the cellar and Mandy said "all right then, come on".' 'The telly wouldn't worry him, if I sent him to bed he'd come down.' 'He's not interested in the telly and I don't buy them any sweets.' 'I can't turn the TV off - it would spite the others. If I had two rooms, I'd do it.'

The percentage of mothers who are prepared to tell the child that they no longer love him is 23, many of these mothers are not certain, however, that the child takes notice. 'He says "I'm not worried".' Others report that it upsets the child. Some mothers insisted that it makes no sense in a large family, 'I have no favourites.' Seventeen per cent of middle-class and 42 per cent of working-class mothers are reported as using this technique in Nottingham.

Do you ever say that you'll send him away or that you'll have to go away from him, if he is naughty? Do you ever try and frighten him with somebody else - his daddy or a policeman or a teacher or a doctor - someone like that? As shown in Table 10.3 the use of idle threats is significantly associated with social class in Nottingham, and our findings confirm this. 'You wait, I'll tell your father' or 'Father's coming' when he was nowhere to be seen was frequently uttered in our presence. One mother, whose husband was temporarily absent added that the kids did not 'play up so much when there's a man in the house'. The policeman is used by some families: 'If Leslie runs off I say "If I see a policeman now, I'm going to take you straight to him." I don't threaten them very often with their father, although I have done with Stephen.' Only two mothers use 'fear' figures: 'I say I go to get the gypsies, it frightens her to death - if she plays up or spits and shouts I say I fetch the firemen.' 'There's a feeder for the railway (i.e., a water tank) beyond the end of the yard, children have been drowned in it. I tell them there are rats and a monster in there, "It'll come up and drag you down", well, you have to say something to them.'

The majority of mothers (61 per cent) said they did not use threats: 'I don't think it's right to frighten them.' 'She's nervous.' 'It would play on his mind.' 'You never know when you need the police.' 'You have a bad time when the child *has* to go away to hospital.' 'I did once say I'd leave her but it didn't get me anywhere, she told me to go on then.' 'He says "I'm not worried", he ain't very bothered about his Dad and he doesn't mind the police.' The mothers who admitted using threats often qualified their statements: 'I'm often saying that, they are frightened of that, but I think it's daft in a way, because if they went to hospital then they might think I'd sent them away on purpose.' 'I say "I'll have you put in a home", or I say "I'll leave you", but I have been in hospital that many times, that doesn't work either - I get the next-door neighbour to shout at them.'

TABLE 10.3 Threats to 3-4 year olds (percentages)

Class:	Nottingham						Socially handicapped families
	I&II	IIIn-m	IIIm	IV	V	All	
Authority figures	10	23	29	29	46	27	27
Fear figures	1	1	2	1	1	1	4
Send child away	8	18	18	19	26	17	20
Will leave child	4	18	14	17	9	13	16

The older boys

The use of threats, we assumed, would be practised by some mothers and not by others, with little discrimination of the two younger age groups. We could not make the same assumption in relation to the 10-11 year old boys. Mothers first commented on the relative uselessness of depriving the boys of television viewing, or other forms of punishment. 'If I put them to bed it doesn't work, I'm soft, I fetch them down, I let them back down to watch the film on the telly.' 'I don't stop sweets - they are not that bad - though I say I'm going to.' 'I think it's best to leave him alone.' 'I shove him out the bloody door and he gets himself back to normal.' A sizeable group of mothers (nineteen in all) frankly admitted using threats, but as a manner of speech rather than a method of frightening the boys. 'I do thousands of times - I'm leaving him every day. They say, "When are you leaving us, Mum?" as a standard joke.' 'I've threatened to leave him but he couldn't care less.' 'They say, "you can go, we've got our Dad".' 'If I say "I'm going", he says "Well then, go" - he isn't bothered.'

During interviews we recorded some of the methods by which mothers or fathers disciplined their children. The following is a descriptive account of three situations in which threats were used frequently. In all, nineteen families had developed an 'idle-threat' technique of this type.

1 Philip, aged six, appeared in the kitchen striking matches. Mother drew Father's attention to it. Father: 'Well, hide the bloody things.' Philip now turns to a younger sister who is playing with a polythene bag, he grabs it. Mother calls 'Robert' and gets no reaction from her husband. Then to Philip: 'You won't get no potatoes tonight.' A little later Philip is heard muttering 'pig bum, apple bum', which causes mother to say 'If you call me that again, I'll spank you.' He went on muttering, but mother took no action.

2 Karen, aged six, cried over some hurt inflicted on her by a sister. Mother called 'Stop snivelling, baby', but the noise and quarrelling went on. Mother: 'Pack it in or you get sent to bed with no tea.' This threat was repeated more than once, there were short silences and sullen looks, then the noises gradually started again. When the girls continued quarrelling, mother said: 'I'm starting smacking', but she did not, and the girls went on. Several further incidents induced mother to threaten smacking again, and after one of these she turned to the interviewer and said 'They don't take no notice.'

3 The children had been sent upstairs to bed, and there were shouts and disturbances. Mother went to the foot of the stairs and shouted 'Julie, I'm going to hit you if you're out of bed.' She then went upstairs and there were yelps and bumps. As she came down she shouted 'I'm going to tell this lady from the welfare to put you away.' There were more noises from upstairs, and mother went to the foot of the stairs again: 'Julie - wait till this lady has gone, you'll wish you'd never had me, I'm not playing, there will be no money for sweets - I'm warning you.'

PROBLEMS OF SCHOOLING

Parental contact

How did parents perceive their roles as third partner in the educational process? The majority had no views on the objectives of education; they were largely accepting, though not necessarily supporting, the educational system. Teachers were seen by many as authority figures whom they are prepared to back up in conflict situations. The majority of parents felt welcome when they visited school on parents' day. But at the same time less than one-third were satisfied with their somewhat tenuous relationships, and many commented on the absence of real dialogue: 'I haven't been as often as I'd like - anyway, you only see their work.' 'Teachers haven't the time.' 'They ask you not to go inside the gates, they might send for you if they are really worried.' 'When I went up I was wandering about like a fool, I didn't know where Wayne's work was.' 'I don't think they want much to do with the parents, I don't think they bother very much. Mind you, they have quite a lot of problem children - it's 90 per cent immigrants as far as I can see - they have their hands full.' 'If father is out of work they make you feel low.' A small number of mothers had no contact with school: 'I never bother to go, I've been invited, but I never bother to go.' 'I never see them, I don't particularly want to.' These comments express lack of self-confidence and uncertainty about the parental role. Contact with school was made in critical situations, initiated either by the parent or by a teacher asking the parent to visit. The percentage who had established contact with school was rated by teachers as 29 per cent for 6-7 year olds, and 20 per cent for 10-11 year olds. The incidents that caused visits were behavioural problems, or complaints about unfair treatment. Only two parents of the older age group of boys had consulted their teachers on the boys' academic progress.

Twenty-eight boys aged 6-7

Does he like school? All but three mothers said their boys liked school. This frequency of positive responses is in line with Nottingham families (86 per cent, no class differences). On being asked what the boys enjoyed best more than half had to admit that they did not know, the others mentioned painting and drawing, writing and reading, or just playing. But there were lively comments when they were asked about dislikes: 'Being confined in the classroom.' 'He doesn't like his exercises - I think he's shy undressing - I say "you've got to do your exercises" - he doesn't eat his pudding, I don't say to him he's got to eat up, I wouldn't force a child to eat up.' 'He doesn't get enough dinner.' 'The other children hit him.' 'He says the teacher hits him, but when we go down, she hasn't hit him - he thinks we'll keep him away, he's trying to get us to keep him away from school.' 'He has one teacher he didn't like, she hit him with a ruler, John says he doesn't know what for.' 'He complained only once when children set about him.' 'He doesn't like being by coloured kids - they smell.' 'He didn't like school at first, he kept crying, he took about four to five months to settle in - he wanted to stay home - he said he was being bashed.'

How do you feel about children's complaints of school? Do you think you should take them seriously? Over 80 per cent of Nottingham professional families take their boys' complaints seriously in contrast to about 60 per cent of working-class families. The percentage in our sample was 50. The mothers who do, indicated they would take action if there were complaints about severe punishment, or about being bullied by other children. 'I've been round once for a reason - he said a child kept duffing him up. The teacher agreed it had happened and she'll speak to the child. I don't like to see them really crying if they are bullied - I'd go mad.' 'When Chris's teacher smacked him for being away, just before Christmas he was off for two weeks with bronchitis. Father went to see the Head, and he said "I'll see the teacher" - the Head refused to let Father see the teacher, he said he'd look into it. The teacher has said nothing to Chris since.' The mothers who do not take these complaints seriously commented that children 'make things up' or 'get things very exaggerated', or that their complaints were unjustified: 'Rosaleen is the worst we've had. She said the lads in her class were ganging on her, the teacher said they don't. She was caned for truancy, she's a difficult child even at home, self-willed.' 'He said teacher smacked him for telling lies - I said, he deserved it.'

Does he ever say he does not want to go to school today? What do you do? Eleven mothers (39 per cent) said he often or sometimes did: 'If we've had a row.' 'Only when the black kids hit him.' 'Only when he's been away for a time.' 'When he had that teacher.' 'He only says it to get threepence.' 'I think it's more if his brother stops home if he's not well.' 'He often comes out with some excuse like a bad ankle.' This frequency distribution is in line with Nottingham professional classes (37 per cent); the percentage for working-class families is 53. Asked what they did on such occasions, five of the eleven mothers are not sympathetic: 'I'd bring him'; 'I say he's got to go'; 'he goes - I don't worry about what he says'. All but one of these boys have normal school-attendance records. Three mothers show sympathy but make the boy go: 'I ask him if anything is the matter, and if he says no, I make him go.' 'I find out what's troubling him.' 'He always quietens down after a few days.' These three boys have bad attendance records. Three mothers accede to their son's demands to stay home: 'It's not very often (that the black kids him him) he still goes the next morning.' 'I just keep him away, I don't argue with him.' 'If I feel like it I keep him at home, if I feel he's not happy then I keep him - if he *does* go then I feel unhappy - I go over at dinner time and have a peep to see if he's all right.' Two of these three boys have bad attendance records.

Does he ever pretend he is not well so as to stay home from school? Twenty-three per cent of Nottingham professional families stated that their sons often, or sometimes, malingered; the percentage for skilled manual workers' families is 32; for semi-skilled 43; and for unskilled 22. Nine mothers (32 per cent) in our study admitted their boys tried to malinger. Of the eleven boys who sometimes say they do not want to go to school, six use the excuse that they do not feel well. Three other boys are labelled malingerers by their mothers: 'He grumbled about a sore stomach, I kept him home. If he thought of doing it again, I'd send him - that sort of thing is not to be listened to - I wouldn't take it seriously.' 'Three weeks ago he strained his back - he played up at school, and the teacher brought him home. Aunt and uncle were still here, so he wanted to be at home, he stopped at home the rest of the week.' The mother of a chronic truant told us 'He does pretend he isn't well ... I'd make sure he wasn't ill and then I'd make him.' Two mothers whose sons have bad attendance records admitted: 'He often says this - I just leave him,' and 'I don't know - I'd see what happened, I think I'd let him stay home.'

School attendance over two terms was ascertained from the school registers. The boys were divided into five groups:
1 normal attenders, 90 per cent of more attendance;
2 legitimate absence, 75-90 per cent;
3 legitimate absence, less than 75 per cent;
4 suspect absence, 75-90 per cent;
5 suspect absence, less than 75 per cent.
Of the twenty-eight boys under discussion seven fall into group (4) and ten into group (5). Only nine of these seventeen bad attenders were said by their mothers to use excuses for not going to school: we have no information about the other eight. They may truant without their mothers' knowledge, or they may be kept home for reasons other than their inclinations, or mothers may not have wished to discuss the matter.

Twenty-nine boys aged 10-11

We would like to know something about how he is getting on at school - does he seem happy? Again, the majority of mothers maintained that the boys were happy, or happy enough, at school; only six thought the boys disliked school: 'He hates the sight of it - he's only been like it since we've moved here.' 'He doesn't seem to have any interest in school work.' 'Not very much - I don't know why.' 'He hates it - for no particular reason.' 'I can't say why.' The other mothers: 'He's settled down well – he doesn't like losing time off school' (bad attender). 'He wasn't keen on the last teacher but now he is' (bad attender). 'He's always anxious to get there – at eight o'clock he wants to go there, he hates to miss a day, he must be happy. Yet I don't think the teacher likes Vincent, he's never got a good word for him, but that doesn't put Vincent off' (bad attender). 'He doesn't complain' (good attender). 'Tony loves school, it's not hard to get him up in the morning' (good attender). There is clearly no correlation between attendance at school and the statements made by mothers about their sons' feelings. School attendance (as defined on p. 157) was bad (group 4) in the case of four boys, and very bad (group 5) in the case of thirteen boys, yet only four of these seventeen bad attenders were said to be unhappy at school. The theme was taken further: *On the whole have the teachers been understanding and sympathetic to him?* Twenty mothers (69 per cent) believed the teachers are understanding people, some mentioned specific instances such as sending Christmas greetings or presents to a boy while in hospital. Nine mothers maintained that the teachers had not been sympathetic; the answers can be grouped under two headings, those that had a particular grievance, and those that were expressions of negative feelings about school in general: 'Patrick was going through a stage of lifting money, and I went to school to see if they would have a word with him, but instead of having a talk with him, they more or less scolded him.' 'I can't send him when he's got his asthma – he's not at school regularly and they don't seem to know his case.' 'The teachers used to find fault ... they insisted he brings so much money for different funds, I can't afford it – I think they take it out on the children who don't bring it.' 'They don't learn them anything, no schools round here are any good – they are all cheeky. They don't learn them enough, they don't get a hiding if they've done wrong. They don't care, they are only good for "cantin" – they spend more time drawing than learning.' These are the mothers of good attenders. In contrast, the mothers of three bad attenders: 'Karl's been off for a long time, he was off with his abscess, and because of me being ill. I don't think he's very bright, he doesn't talk much about school, he's not interested. I find they don't help you – say, if your child has no clothes or shoes – I've had little help from anywhere.' 'I think he dislikes the service in hall, he hates anything religious, I hate it.' 'Vincent is not interested in school work, he's got much imagination – his mind drifts. He used to make up stories, he still does. I've had to go to school because of the teacher picking on him.' Twenty mothers (69 per cent) spoke warmly about the teachers: 'He lets them do things they are interested in.' 'Vincent always takes the teacher's part.' 'He likes some of the teachers.' But nine of these boys are bad attenders. There is clearly no correlation between regularity of school attendance and mothers' perceptions of their sons' feelings about school.

Has he ever been in real trouble with school so that you really had to speak to him seriously? Only two mothers said they did: 'It was a case of having to talk to him a lot – he was going through an emotional upset, he thought we didn't care for him.' This mother had advice at a child guidance clinic after an episode of stealing at school. The other mother: 'Only for fighting – Paul won't be picked on.' *Has he ever played truant from school as far as you know?* Thirteen mothers said their sons had truanted: 'Only once with a boy I stopped him going with.' 'When he was younger - Linda started and took Paul with her, it went on for about a month before they were found out.' 'Yes, he takes off days here and there - the school board man told me.' 'You are putting it mildly! If he doesn't go to school he doesn't run off, he stays here.' This mother expressed much anxiety about her son's non-attendance. 'He has done once or twice, the teacher and Charles caught him, father smacked him - it didn't make any difference.' 'Somebody I knew saw him - I threatened to tell the Head and I don't think he's done it since.' Sixteen mothers said their sons did not truant; eight of these boys have bad attendance records. However, non-attendance for suspect reasons is not necessarily truancy. Two parents gave an account of their difficulties in getting the boys fitted out for school and the humiliations their sons suffered which led to refusal to go to school.

Do you think teachers should be allowed to use the cane or strap on children of his age? Has he had the cane or strap at school? Did you think it was right that it should have been used? Opinions varied considerably. Fifteen mothers thought that teachers should use the cane, but some of them qualified the conditions. 'I know what children are like - I've been caned at school.' 'I wouldn't complain, but I'd want to know why.' 'I've known him to have had a smack with the ruler, I told him he must have been doing something wrong - I don't think they do extreme damage.' 'Teachers should use the cane if they are really naughty.' 'Only for something really bad, not for fighting or ordinary things.' Fourteen mothers were against the use of the cane: '*I* don't use anything - why should they? There are other ways of correcting them.' 'He's had quite a lot of it at first - it started him that he didn't want to go to school. (*Did you complain?*) I did but they said they were the teachers, what happened was their responsibility.' 'Children are at school to be taught, but not to be hit.' 'If he came home marked, I'd go and complain.' Nine mothers knew the boys had had the cane, twelve believed they never had the cane, and eight were not sure.

Some people think you should always back up the teacher to the child, how do you feel about that? The mothers who did not criticize the teachers for using the cane were inclined to back them up when the boy complained; eleven of them emphatically so. 'I take the teacher's point of view, they are more qualified. Take Michael - he grumbled one day that the teacher had "murdered" him. I went up to the school, he had exaggerated, he'd got in a tantrum and wouldn't sit down.' 'Yes, I think you should - when they are in school they are in the teacher's care, I back the teachers up.' 'I tell the teacher if they are cheeky and telling lies to smack them.' 'I won't dress teacher down in front of the child, unless there's real victimization.' 'He takes more notice of his teacher than of others.' Another eleven women thought the teachers should be backed up in general, but they would not like to see the boy severely punished: 'If they belted and marked them for something trivial, then I'd go down.' 'I always say if they have been hit they must deserve it. If they had weals on them, I would complain.' 'I back him up if the children are in the wrong, but I don't like teacher hitting him behind the ears.' 'You send them to school to be taught, they should be corrected in the proper manner. If they came home with a mark on them, I'd go down and see what was up - not for every time they got a caning though.' 'I'd have to hear both sides of the story.' 'I'd back the teacher up if the child is in the wrong, but a child can't be in the wrong all the time. More often than not I'd take the teacher's part. Vincent is a terror at home, but the way the teacher spoke to me on the Open Day - he hadn't one good word to say about Vincent. No child has all bad points, it hurts me to think he hadn't one good thing to say about him.' Only seven mothers (24 per cent) expressed the view that teachers are no different from other people and don't need backing up: 'They have too much authority as it is.' A father and mother put it this way: Father: 'I wouldn't always back up the teacher.' Mother: 'I'd listen to the child first to find out if he's genuine, then I'd go to the teacher and hear his side and weigh it up. The trouble with a lot of parents is they aren't strict enough.' Father: 'That's where they make a mistake, they've stopped the cane. In my opinion, how is a child going to learn if not corrected properly? The cane never hurt anyone.' This mother told us she never smacks, but that father does, but he does not use an instrument. Another mother: 'It's all according to who the teacher is - some are very strict. Peter wet and soiled for months because he thought he was going in her class, Denise was afraid of her too.'

In summary, the families in our study resembled working-class families in Nottingham in that three-quarters smack frequently although a number of these did not think that smacking is an effective form of sanction. Mothers tend to smack more often when noisy and quarrelsome behaviour irritates them, and less for disobedient behaviour than their counterparts in Nottingham. Again, as with Nottingham working-class families, the use of language is not found to be effective in overcrowded conditions, but threats are uttered to ventilate feelings, and in some families children are scared by them. The regime of the home is closely related to the temperaments of the children. Mothers tended to take their sons' complaints about school rather less seriously than Nottingham mothers, and there was a variety of reactions; but maternal practices did not correlate with the boys' attendance patterns. Comments made by parents of the boys aged 10-11 showed that the majority of parents (76 per cent) accept the teacher as a person of authority whom they are prepared to back up against the child in conflict situations. In that sense they continue to play a traditional role. But there were others who were highly critical, who tended to take up a defensive stance on behalf of their sons, and who were prepared to back the boys up in all and any of their wishes when these conflicted with the demands that school made. Parental comments about the boys' dissatisfactions and complaints at school were concerned with personal relationships and with stressful situations and not with academic properties.

Part three

INTERPRETATION

Chapter 11

CHILD-CENTREDNESS

The preceding chapters give evidence of child-rearing practices which are on the whole in line with that section of the Nottingham sample which belongs to working-class families, and when there were linear class differences our sample tended to resemble the non-skilled sector. There are, however, important divergences. In training their children for independence our sample was found to be constrained by limited resources. Lack of money prevents children from taking part in activities which are enjoyed by their counterparts in Nottingham. Membership of large, low-income families means less individual attention; it also means delegation of essential maternal functions to older siblings. It means training the very young child to stand on his own feet in encounters with other children and in ensuing quarrels. Residence in crowded accommodation, often in terraced housing and often around communal yards, imposes obligations towards neighbours which take priority over the needs of the children if harmony is to be preserved. There are considerable differences in the children's experiences of personal possessions; over half of Nottingham's non-skilled families are able to provide toys for individual children; the number in our sample was negligible and toys are shared. It was difficult to compare mothers' participation in their children's games since character and quality of play differs between the two samples. It was equally difficult to evaluate mothers' attitudes towards the young child's need for undisturbed play, since the deprivation of toys and absence of privacy for play prevents the child from becoming absorbed in play. The handling of 'pocket money' was found to differ to such an extent that it proved to be impossible to compare our findings with those of the Nottingham sample. Money, like other possessions, is shared and not allotted to individuals, and there are neither ownership rights nor expectations of regularity of supply. It was not possible to compare the sharing of special interests between parents and their older sons, again because the parents were unable to pursue hobbies for lack of money. Family holidays, outings to theatre, concerts or museums are unknown to almost all the families in our sample. These and many other differences in life styles between our sample and the Nottingham sample highlight the extreme constraints put upon the activities of families who live in poverty.

Poverty, as defined in this study, is measured on a scale used by the Supplementary Benefit Commission; it provides a minimum income below which it is considered undesirable by Parliament that any family should have to live. In fact, over one-third of the families in our sample fell below this minimum standard. Fifty-two per cent were found to have incomes of the size normally covered by supplementary benefit, and only seven families had incomes exceeding this narrow poverty band. The experience of poverty was common to all families at some time, whether in or out of work. Linked with poverty is residence in low-rent accommodation which for almost all families meant lack of basic amenities, like indoor lavatories, a hot-water system, a bath, or bedroom-heating facilities. On the basis of a bedroom standard, 79 per cent of the families were found to be overcrowded, some severely so. All these material factors affect the life styles of the families. But a number of families had to bear additional stresses; these were disability of either parent, an invalid child requiring much nursing, an exceptionally large family consisting of eight or more children of whom three or more were under 5 years old, or financial harrassment suffered in consequence of continued existence on an income beneath the poverty line. Only twenty families were not subjected to any of these additional stresses, twenty-four were subjected to one, and twelve families to two or more. None of the homes in the latter category were found to be happy. While the degree of happiness did not entirely depend on mothers' self-rated ability to cope, but also on parental marital relations, there was a noticeable association of a happy home atmosphere and the absence of additional stresses.

Many of the child-rearing practices in Nottingham were found to be related to social class. Traditional working-class practices are described as tending to be rather authoritarian, and middle-class practices as tending to be more democratic in the sense that the child is seen as an individual with rights of his own which must be respected. Although for a number of practices the differences between professional-class mothers and non-skilled, working-class mothers are of statistical significance, many of the responses given show a wide range of attitudes in all classes. Class differences are sometimes diminished by generational changes; a number of working-class parents are recorded as desiring a closer relationship with their own children than they had experienced with their parents. The dimension which is explored in these relationships between parents and their children is widened by adoption of the concept 'child-centredness'. The Newsons describe this concept in the following manner: 'Other workers in the field of child-rearing have proposed the dimension of child-centredness as a measure of some importance in assessing methods of upbringing, and we would agree that, while it has much in common with the authoritarian-democratic dimension, it adds something to it in terms of warmth and a willingness to take trouble for the attainment of aims in which the parent has no immediate interest' (1968, p. 489).

A composite index was devised by the Newsons consisting of items from the interview schedules to bring together information indicating

'child-centredness' in the mother. The following items were chosen for the 4 year old:
*1 mother is 'invariably' or 'mostly' responsive to the child's demands;
 2 mother participates wholeheartedly in the child's play;
*3 mother is willing to chat during the night at the child's request;
 4 mother unreservedly accepts child's 'I'm busy' excuse;
 5 mother values child 'for himself';
 6 mother chooses not to make a moral issue of the child smacking her.

Scores were allotted on a scale giving one for each item in which the mother is positively rated. Significant class differences were found to exist on comparing low scorers (0-2) with high scorers (4-6), middle-class mothers having 25 per cent of low scorers and 43 per cent of high scorers, in contrast with working-class mothers with 38 per cent low, and 32 per cent high scorers.

For the age group of 7 year olds the following items make up the child-centredness index:
 1 he has a special place of his own where he can keep his own things;
*2 his friends come and play at his house 'most weeks';
 3 if he doesn't want to go to school, mother shows sympathetic attitude (whether child goes is irrelevant to score);
 4 if he pretends he is not well so as to stay at home, mother 'allows off' or 'shows sympathy but makes him go';
 5 mother takes his complaints about school seriously (irrespective of action);
*6 mother keeps or displays at least some of his drawings when he has finished them;
 7 mother shares a special interest with him;
*8 when deciding what to do during holidays the boy has a say in what parents choose;
*9 when he is rude to mother she merely rebukes or does nothing (i.e., does not punish in absence of discussion);
*10 mother will say 'sorry' to him for being cross (apology because of mistaken accusation excluded).

Class differences for this age group were even greater; middle-class mothers had 15 per cent low scorers (0-4) and 51 per cent high scorers (7-10) in contrast to working-class mothers with 40 per cent low, and 25 per cent high scorers. An index of child-centredness for the 11 year olds was not available at the time of our fieldwork.

In chapter 8 (pp. 107-8) the reasons for curtailing the Nottingham interview schedules were discussed. Two questions relevant to child-centredness towards the 3-4 year old, and five questions about the 6-7 year old boys were among those that had to be omitted. The items in the two above indices which we were unable to score are marked by an asterisk. This made an identification of 'child-centred' mothers in our sample impossible; yet on the basis of the remaining items we feel confident to discuss the validity of this measure as applied in circumstances of poverty. The milieu of poverty imposes environmental and situational constraints which severely limit choices. If there are no alternatives and methods of child-rearing are determined by expediency, or if the preconditions leading to the situation which is being examined are very different from those taken for granted by the observer, the result of applying the measure may be quite misleading.

The following observations refer to the index for the 4 year old. Maternal functions are frequently delegated to older siblings, especially at night since the children usually share their bedrooms, and therefore the degree of her responsiveness and availability at night are irrelevant. In homes where children rarely or never engage in intensive play the mother's *wholehearted* participation' in play cannot be expected. Similarly, the mother's attitude to the child's right to undisturbed play cannot be ascertained. Mother's appreciation of the child's personality involves considerable linguistic skills. The index for the 7 year old presents further problems. 'Sympathetic understanding' of the boy's reluctance to go to school could well be no more than an expression of lack of concern; it is 'child-centred' only if normally the mother insists on regular attendance. Mothers who do not take their sons' complaints about school seriously should not necessarily score low on the index, as their attitude may be an expression of a feeling of incompetence, a loss to know what to do about these complaints; or they may realize that their sons' problems arise out of the clash between life at home and demands made at school which the boys cannot meet, and which is beyond their understanding. Many parents in large families do not share *special* interests with their children because they lack the means to pursue special interests. The measurement of child-centredness as defined in the two above indices is inappropriate in the milieu of poverty.

WORDS AS AGENTS OF TRUTH

Child-centredness as a measurable attitude is complemented and more fully elaborated by the Newsons in discussing mothers' use of language when faced with behavioural shortcomings. Many mothers, especially in classes I and II, attempt to convey to the child the distinction between 'maternal love which persists under all circumstances' (1968, p. 497) and their sometimes punitive reactions to behaviour that they cannot tolerate. The use of language becomes of great importance. The class differences found by the Newsons in the kind of sanctions that are used exist also in the methods of reasoning or the use of threats. The meanings that words carry as explanations and as predictions become of central importance in mother-child communications. Class differences reach significance levels in all these dimensions. Middle-class mothers in Nottingham interact with their four year olds more frequently and in more verbally articulate ways than working-class mothers. A higher percentage arbitrate in children's quarrels, thus giving them a basic understanding of justice and the principles involved in acceptable social behaviour. In specific situations middle-class mothers more frequently reason with their children, and working-class mothers, on the whole, prefer to use 'short-cut methods which get results quickly' (1968, p. 434). Some of the information analysed in the Nottingham study could not be ascertained in ours, as interview schedules had to be shortened, and comparative material on talking at the dinner table, bed-time stories and prayers is not

available. But as far as comparison is possible, our families tend
to show a frequency distribution of responses that is in step with
Nottingham working-class families. Our families on the whole also
find smacking for naughtiness the most effective punishment, and
about half the group will threaten smacking or smack the child for
lies or for temper tantrums, whereas less than one-third of Nottingham middle-class families will do so.

The Newsons also found significant class differences in the use
of unfulfilled punishment threats in dealing with the 4 year old
such as telling the child that some person of authority will intervene, or that the child will be sent away, or that mother will leave
him. Only 10 per cent of middle-class mothers will use authority
figures, in contrast to 29 per cent in classes III and IV, and 46
per cent in class V. An extension of the language factor is used
in the degree to which mothers 'evade or distort the truth' to
demonstrate the importance of this factor in socialization. The
index measuring 'evasion or distortion of truth' includes:
1 a false account of where babies come from;
2 idle threats of authority figures;
3 threats to leave the child or send him away;
4 slipping off without telling the child, leaving him with a minder;
5 frequent use of unfulfilled punishment threats as judged on the
  basis of mothers' general discussion of the way in which they
  dealt with discipline;
6 any other definite instance reported by the mother or observed by
  the interviewer.

Mothers who scored two or more on these six items were described as
'more evasive'. The Newsons put much emphasis on the valuation that
is placed upon words 'as the agents of truth'. The professional-
class mother will 'take great care not to say anything to the child
which might be considered untruthful', but as one descends the social
scale 'the general attitude towards what is or is not permissible
to say to the child becomes far more a matter of experience and quick
results'. Threats of 'unenforceable sanctions' are a distortion of
truth, since 'discipline is in fact maintained by verbal means, in
which the mother, so far from being at a loss for words, deliberately
uses the more subtle power of words to suggest more dramatic sanctions
than the simple slap, in order to attain her immediate ends' (1968,
pp. 470-2).

The division of mothers into those who are 'more evasive' and
those who are 'less evasive' on the basis of the above six criteria
appears to us to be inappropriate when applied in circumstances
other than those of the typical small, middle-class family. Item (1),
a false account of the birth of babies, is a method widely used to
veil the realities of procreation in an endeavour to protect the
small child's sexual naivety. In many working-class families such
a method is governed by customary practices, and it need not indicate
a general tendency to distort the truth in other situations. Item
(4) 'slipping out without telling the child' may attain significance
in a family in which close and continuous mother-child contact is
typical. In such a setting the sudden disappearance of the mother
without warning may indeed be an 'evasion of truth' appearing to the
child as a betrayal of trust. The situation is very different in
large families, or in three-generation families, or in families who

employ staff as child minders; in such a setting mothers will typically see little reason in giving account to their children of all their movements.

This was in fact the tenor of many responses in our study: 'I could tell them, but they take no notice.' 'I would tell her if she asked.' 'They don't bother much.' Two-thirds of the sample indicated clearly that there was no 'deception' in this kind of situation. However, nineteen mothers (34 per cent) make a point of going shopping without 'broadcasting the fact', primarily because it avoided a 'scene': 'It's better to slip out when I'm going shopping, Lisa would scream and scream, Brian used to be the same until he went to your play school.' 'I wouldn't tell the little ones, they all seem to be after me.' 'The trouble is Marie, if she spots me, that's that she even follows me to the toilet.' Obviously the 'clinging child' is a phenomenon even in the large family, but not in many. Mothers were more concerned about the nuisance factor in trailing several small children around the shops, and the children appeared to be more upset about missing a treat than being abandoned by their mothers.

The distribution of 'more evasive mothers' on the Nottingham index is as follows: class I and II, 12 per cent; class III non-manual, 29 per cent; class III manual, 42 per cent; class IV, 37 per cent; class V, 62 per cent. In spite of our misgivings we have, for comparative purposes, scored the responses of mothers in our study and found 52 per cent to be 'more evasive'. However, if the two items 'false account of birth of baby' and 'slipping out without informing the children' are removed from the index, and the remaining items are counted, the number of 'evasive mothers' is reduced to 21 per cent.

The index is concerned to measure attitudes towards the 4 year old. Many mothers told us that they do not think the use of idle threats to be appropriate with children of that age; but they will use such threats with older children. Nineteen mothers admitted this, but their statements were frequently qualified by observations about the ineffectiveness of threats. 'They say "when are you leaving Mum?" as a standard joke' describes the feelings of these mothers. One enters here into an area of forms of human communication in which analysis of the meaning of words may not be entirely in place. The use of idle threats as a form of expletive, indicating the feelings of the speaker, is widespread, and in families where this is habitual it is probably taken as an indication of the speaker's mood rather than as words symbolical of truth which are meant to deceive.

AN ALTERNATIVE MEASURE OF CHILD-CENTREDNESS

The two Nottingham measures indicating degree of 'child-centredness' and degree to which mothers 'evade or distort the truth' were subject to misinterpretation when applied to families in our sample. An alternative way of ascertaining the quality described as 'willingness to take trouble for the attainment of aims in which the parent has no immediate interest' appeared to us possible, since the existence of such willingness was evident. There were two family characteristics which merited attention; one was the degree of parental participation in their children's activities, the second was the trouble taken over supervision of the school boys, in other words the degree of chaperonage exercised. Which of the two, if they were independent of each other, was the better measure? The first task consisted of identifying families under each of the two headings; if indeed the families who participated a great deal in their children's activities were also the families who exercised a high degree of chaperonage the problem of defining 'child-centredness' in the milieu of poverty would be solved.

The degree of parental participation was discussed in chapter eight. While mothers' 'wholehearted participation' as defined by the Newsons took place in only 24 per cent of the families in our sample, on a less stringent definition we found 54 per cent who habitually play with their 3-4 year olds. Again, while very few fathers were found to share 'a special interest' with the 7-10 year old sons, we identified forty fathers (71 per cent) who 'liked doing things with the children'. Participation on this level was 'high' in nineteen families, and 'fair' in twenty-one families. Fathers' participation in the management of the 3-4 year olds (in accordance with the Newson definition) was 'high' in nineteen and 'fair' in twenty-four families (p. 120). The measure called 'chaperonage' was described and discussed in chapter nine. 'Much' chaperonage was exercised in sixteen, and 'some' in twenty-five families, but fifteen families exercised no such measure. Was there, we wondered, an association between 'participation' and 'chaperonage'? As shown in Table 11.1 there is only a chance relation.

TABLE 11.1 Chaperonage and parental participation

	Chaperonage			
	Much	some	none	Total
Mothers and 3-4 year olds				
- plays	8	12	10	30
- does not	8	13	5	26
Fathers and 3-4 year olds				
- high	6	7	6	19
- fair	7	13	4	24
- little or none	3	5	5	13
Fathers and 7 or 11 year old boys				
- high	9	4	6	19
- fair	3	14	4	21
- none	4	7	5	16

(Mothers: Chi-squared = 1.42 with 2 df, chance expectation. Fathers: 3-4 year olds: Chi-squared = 2.80 with 4 df, chance expectation; older boys: Chi-squared = 9.34 with 4 df, not significant.)

The degree of parental participation in the management and activities of their children is fairly evenly divided between families who exercise much, some, or no chaperonage. Obviously the two measures characterized different kinds of families, and they had to be treated as independent variables.

Taking the families who showed a high or fair degree of participation in the management and activities of their children we had earlier established a cluster of other characteristics. We found a highly significant association with home atmosphere (p.121), and the happiness of the home was significantly related to absence of additional stress (p. 101). There was, in the homes where families had fewer burdens to carry, a happier atmosphere, and this was associated with a greater readiness of parents to join in the activities of their children. In addition parental participation was positively associated with the existence of toys and equipment suitable for play (p. 110), and with parental hobbies (p. 119). Here was obviously a group of families who provided a more varied and, in terms of human relationships, richer life for their children. In a fair number of these families fathers' participation correlated with the intensity of the school boys' activities, although this did not reach significance (p. 122).

But other child-rearing measures did not correlate with the type of family described above. Neither home atmosphere nor parental participation determined whether there was strict modesty training; whether boys helped regularly in the home; or whether they attended school regularly. However, a cohesive family, a happy home atmosphere, and parental interest and participation in their children's activities have traditionally been considered to be of importance in child development. We had three measures by which we were able to describe the children's functional development: attainments at school, behaviour in the classroom, and delinquency. We have reported a generally poor showing in reading tests of even the abler boys in our sample in chapter 4; and these findings are further discussed in the final chapter. There appears to be no association between doing well on tests and any family characteristics. Similarly, we did not establish an association between fathers' participation and behavioural indicators (as rated by teachers). Again on comparing families in 'happy' and 'unhappy' homes the distribution of behavioural indicators of the school boys was random.

All families were subjected to an examination of delinquency (chapter 6); twenty-seven families were found to have delinquent children and twenty-four families were non-delinquent. It was shown that there was a tendency for the delinquent families to have more children at risk, more boys at risk, and more boys aged thirteen or over than non-delinquent families. As families the former were more at risk. As this observation indicates, the family unit is not a satisfactory measure to examine an association of delinquency and home atmosphere or parental child-rearing methods. To obtain a more accurate measure the number of child-years at risk was established for each family. Of the fifty-one families who had children aged ten or over the range of child-years at risk covered one to fifty-two years. The total number of child-years at risk in the entire sample was 515 for boys, and 345 for girls. (The excess of boys over girls is explained on pp. 77-8). We thus had four measures of delinquency:

1 delinquent and non-delinquent 13 year old boys;
2 delinquent and non-delinquent families;
3 recidivist families (being a group of delinquent families; whose children had offended repeatedly) and other families;
4 boy-years at risk and number of offences.

In comparing these measures of delinquency with home atmosphere and with fathers' participation in the school boys' activities no relationship with either of these two family characteristics could be established. Table 11.2 and Table 11.3 show the findings of a selected group of these measures.

TABLE 11.2 Home atmosphere and delinquency

	Home atmosphere			
	Happy	Intermediate	Unhappy	Total
13 year olds				
- non-delinquent	6	7	6	19
- delinquent	3	6	1	10
(Chi-squared = 2.0 with 2 df, chance expectation.)				
Families*				
- non-delinquent	9	10	5	24
- delinquent	7	13	7	27
(Chi-squared = 0.9 with 2 df, chance expectation.)				
Families*				
- recidivist	5	5	2	12
- the rest	11	18	10	39
(Chi-squared = 0.8 with 2 df, chance expectation.)				

* Five families whose children are under 10 years old are omitted.

TABLE 11.3 Father's participation (aged 10-11) and delinquency

	Participation			
	High	Fair	None	Total
13 year olds				
- non-delinquent	6	6	7	19
- delinquent	4	3	3	10
(Chi-squared = 0.2 with 2 df, chance expectation.)				

Table 11.2 shows convincingly that the home atmosphere is unrelated to delinquent behaviour of the children; in a total of sixteen happy homes seven have delinquent children, and in a total of twelve unhappy homes there are five with no delinquents. Of the ten boys who,

by the age of thirteen, had become delinquent more are found in happy than in unhappy homes. The same is true of the twelve families with a number of children who have offended repeatedly; five of these families have happy homes and two have unhappy homes. Table 11.3 shows that there is no association between fathers' participation in their sons' activities and delinquency; the number of delinquents is evenly distributed among families whose fathers 'like doing things' with the boys, and others who have no interest in them.

If neither of these factors is related to delinquency, specific child-rearing methods may provide the explanation of the sharp contrast between twenty-seven delinquent and twenty-four non-delinquent families. Common sense suggests a difference in parental normative standards of sufficient strength to influence parental measures in handling their children. No good method has yet been developed to ascertain such standards. Questions cannot elicit what is not intended to be revealed. However, the interview schedules designed by the Newsons and adapted for our study elicited certain child-rearing methods which reflect standards obliquely. The relevant set of questions in this context is that contained in the 'chaperonage' index (pp. 145-6). This index measures a number of family practices, some reflecting the degree of independence of the child, others the methods by which parents exercise supervision so as to give the child protection commensurate with his increasing capacity for responsibility. The measure, designed for both age groups of school boys, clearly relates to environment, and it should be evaluated in relation to the nature of the environment. Parents interpret their obligations to protect the child against perceived dangers in many different ways; thus variation of practices can be examined at the level of parental perception of the environment and at the level of parental practices in providing protection.

The environment has been described by quotations from parents in chapter 7 (pp. 89-90). Well over half the families, thirty-six in number, expressed their antagonistic feelings in strong terms. Twenty families were less critical, but many of this group were very conscious of certain negative aspects. The most frequently mentioned aspects were traffic dangers; risks of accidents while playing on demolition sites, near the railway, near the canal or reservoir; the infectious nature of vandalism; hostile neighbours; child-molesters; the temptations of thieving, shoplifting, stripping houses due for demolition of wire and metal; and sex play of youngsters in vacated houses. In our conversations the theme of the 'bad' environment was brought up frequently, and parental quotations are recorded in the context of play, independence training, and control of aggression. In the section on 'undesirable influences' by other boys (pp. 113-14) we reported anxieties about delinquents expressed by a group of parents with 10-11 year old sons. Some of these 'warned' the boys; some explained the likely consequences or merely expressed their disapproval. Others imposed strict regimes of supervision, either by making rules and enforcing them, or by restricting the children's movements, or both. Other parents, preoccupied with hazards other than the contagious nature of delinquency, had imposed similar restrictions. The parents who exercised chaperonage had many motivations, but they shared one characteristic, a strong sense of obligation that their children needed protection against a bad environment.

Were these measures effective? We cannot explore their effectiveness in terms of prevented accidents or interference by paedophiliacs; the only direct measure at our disposal is detected delinquency of the older group of school boys. The assumption in this kind of proof of the effectiveness is the belief that chaperonage exercised when the boys were aged 10-11 affects their behaviour when they are three years older. Table 11.4 shows that this may indeed be the case.

TABLE 11.4 Chaperonage and delinquency: boys aged thirteen

	Chaperonage			
	Much	Some	None	Total
delinquent	1	2	7	10
non-delinquent	6	12	1	19
Total	7	14	8	29

(Probability on the Fisher 'exact' test on 'much' versus 'none' is 0.01.)

The numbers involved in this comparison are small, but the result shows conclusively that there is a significant relationship between chaperonage and detected delinquency three years later; only one boy who had 'much' chaperonage was delinquent aged 13, in contrast to seven who had none. For non-delinquents the relationship was reversed. Since all families were severely socially handicapped in comparison with other residents of the deprived areas of the study, it cannot be maintained that the selective processes by which certain offenders reach the courts and others do not have operated in a discriminatory way.

A second measure consisted of a comparison of delinquent and non-delinquent families. The assumption in this case is more complex. The exercise of chaperonage was established by focusing on either a 7 year old or an 11 year old boy in the family. Many families had older children (see Table 6.1). In comparing child-rearing methods of these two age groups with the incidence of delinquency in the entire family it was assumed that these methods had been applied to other children in the family, and that the effect of these methods had not washed off during adolescence. Table 11.5 shows that this appears to be true.

TABLE 11.5 Chaperonage and delinquency: families

	Chaperonage			
	Much	Some	None	Total
delinquent	4	10	13	27
non-delinquent	12	11	1	24
Total	16	21	14	51*

(Probability on the Fisher 'exact' test on 'much' versus 'none' is 0.0002.)
* Five families whose children are under 10 years old are omitted.

The correlation of chaperonage and absence of delinquency in the family is highly significant. Of the families who operate much chaperonage, three-quarters had remained non-delinquent; whereas of the families who did not operate chaperonage all except one produced delinquents. The highly significant relationship of chaperonage and delinquency at adolescence was confirmed by comparing non-delinquent families with families who had produced several delinquent children, of whom at least one had become a recidivist, as shown in Table 11.6.

TABLE 11.6 Chaperonage and delinquency: recidivist families

| | Chaperonage | | | |
	Much	Some	None	Total
recidivist	2	2	8	12
non-delinquent	12	11	1	24
Total	14	13	9	36

(Probability on the Fisher 'exact' test on 'much' versus 'none' is 0.001.)

In view of the fact that the family as a unit is not an accurate measure for comparative purposes because of variations in size, sex distribution and ages of children, a comparison was made of child-years at risk and number of offences in families exercising much, some and no chaperonage. The results for boys are shown in Table 11.7.

TABLE 11.7 Chaperonage and boy-years at risk

| | Chaperonage | | | |
	Much	Some	None	Total
Years at risk	147	189	179	515
Number of offences	18	19	58	95

(Chi-squared = 29.1 with 2 df, $p < 0.001$.)

(The total number of offences for girls, eighteen, was too small to make a statistical statement.) This comparison demonstrates convincingly that chaperonage has an effect on subsequent offending behaviour. The number of offences expressed as a percentage of years at risk was about one-eighth for boys who had received chaperonage and about one-third for boys who had not received chaperonage. The findings of this and the preceeding tables are sufficiently conclusive to justify a closer examination of a dimension in parenting which lies on a strictness-permissiveness axis.

Chapter 11

PERMISSIVENESS

This concept has been widely used to describe degrees of tolerance towards behaviour which is not appreciated as such, but which for the sake of a more important objective is not eradicated, or not eradicated rapidly. Sears, Maccoby and Levin (1957) considered the degree of maternal permissiveness to be of major importance in relation to all aspects of child-rearing. An attempt was made to relate permissive handling of the child to observable consequences. 'From a theoretical standpoint' the authors admit,

> permissiveness is a puzzling dimension. There have long been two opposing conceptions of the nature of motivation. One views the child as having within him a sort of reservoir of impelling forces that are continuously being replenished... Permissive child-rearing, according to this view is a way of reducing the societal counter-pressures, and hence may be expected to reduce both the strain and tension within and the probability of explosive or neurotic expressions outward.... The other view of motivation is that the main social motives are developed as a product of experience.... From this viewpoint, the child who is allowed to *be* aggressive will be more likely to resort to aggressive actions on future similar occasions. The impelling forces to much of his behaviour, then, are not constantly being replenished, but are activated only when certain cues present themselves to him. And what acts he performs ... will be determined by what ones he has found by practice to be the most rewarding (pp. 486-7).

The authors stated that their findings were ambiguous, and they were unable to provide a conclusion about the nature of motivation and the impact of permissive attitudes by mothers.

The Newsons found the permissiveness/restrictiveness dimension to be 'confusing and inadequate' in comparing the handling of the 4 year old in different socio-economic classes (1968, pp. 434-5). The difficulty lay in the measurement of degrees of permissiveness because of the wide range of activities in environments which were not alike. This difficulty did not arise in our study as the environment restricted activities to a much narrower range. We had as a measuring device a number of child-rearing methods which contained a category of permissiveness. Some of these, however, were hardly discriminatory with the vast majority of parents following the same practice; others applied to selected age groups of children only. We were left with three measures: the index of modesty training (pp. 133 and 135), restrictions on play (p. 110), and mothers' reactions to the school-boys' refusal to help (pp. 135-9). These items formed an index of permissiveness scoring 0-3. The family distribution was as follows: twenty 'strict' (score 0), twenty 'less strict' (score 1), and sixteen 'permissive' (score 2 or 3).

The expectation was that families scoring as 'strict' on these three aspects of child-rearing would also be likely to impose restrictions on the school-boys' movements, or conversely, that families who scored as 'permissive' would not be likely to impose these. Table 11.8 shows that there is indeed a high degree of correlation between chaperonage and the absence of permissiveness.

TABLE 11.8 Chaperonage and permissiveness

	Chaperonage			
	Much	Some	None	Total
Permissiveness score:				
0	9	10	1	20
1	6	8	6	20
2-3	1	7	8	16
Total	16	25	15	56

(Chi-squared = 11.76 with 4 df, p < 0.02. Fisher 'exact' test on corners: p < 0.001.)

In a group of twenty 'strict' families only one did not operate any degree of chaperonage, and in a group of sixteen 'permissive' families only one operated much chaperonage. This significant association of permissive attitudes and the absence of chaperonage carries much additional information about the families in the two categories. Chaperoning families tend to insist on a degree of tidiness and cleanliness in the home, they tend to discourage or punish genital play, children looking at each other when undressed, or giggling over the toilet, and parents tend to avoid undressing in the children's presence. They tend to rebuke or punish the school boy who refuses to help when asked to do so. Conversely, permissive families tend to perceive the deprived neighbourhood as normal or at least as one to which adjustment without special precautions is possible. The school boys are allowed to roam the streets, mothers cannot always locate them, there are no special rules about returning straight home from school or at set times at night, and the boys may have been found wandering by the police.

Other family characteristics do not show correlations. School attendance, contrary to expectation, was not significantly related either to strictness or to chaperonage. Families receiving supervision from the social services department were neither more nor less strict than families who had only short periods of contact with the department; and the same observation held true for the chaperoning families.

The permissiveness index - developed for want of a more sensitive instrument - identifies families quite well in certain areas of family activities. Permissive families appear to have abandoned standards of parenting which the majority of Nottingham working-class parents consider to be of importance. It is one of the characteristics of parenting, in whatever class setting, that behavioural standards are defined, that parental expectations of their children's behaviour are related to these standards, and that there is recognition of achievement and pursuit of failure. What is permitted, and what is considered intolerable varies from family to family, and certain class patterns of variation have been found to exist. Permissiveness, as observed in Nottingham professional families, is typically associated with well-defined conceptions of what child-rearing is about. But the kind of permissiveness we have recorded among a quarter of the families in our sample is of a different nature; it extends over the entire field of socialization. We will let mothers speak: 'I don't force my kids to do nothing.' 'I let them carry on, I stand a lot off them.' 'I keep telling him - he says "I don't want to, I don't have to" - so I do it myself.' 'He doesn't like being told things, I have to more or less bribe him to do anything.' 'If you don't give him what he wants he gives you hell.' 'I soft-soap him, I promise him this and that - he doesn't let you forget.'

Some permissive mothers have quiet, withdrawn children who give little cause for friction, and if parental demands are minimal conflicts rarely arise. Others reported much trouble which they linked with explanations of their temperament: 'He's got a terrible temper, if you cross him he throws things - you've got to get round him.' 'If he can't have his own way he shouts and kicks the furniture - in the finish I give him his own way - I have to.' 'He's the devilment, he does the bloody things you don't want him to do, he's full of everything, he climbs on roofs and out of the windows, he fights his brothers and sisters and kicks at Sunday school. If you hit him he'll go under the bed and scream.' 'If he goes to the shop and wants things he can't have, he just takes them - he seems to think it's there for him to take. I slap him or send him to bed but it doesn't do much good.' 'I've smacked him a bloody lot - it doesn't give any results unless you give him bloody money.' Smacking in these situations is an expression of parental exasperation and not a punitive measure deliberately applied as a sanction.

The families who exercise chaperonage and who tend to adhere to traditional standards of strictness are motivated in many different ways, but they share the belief that the deprived neighbourhood and its inhabitants are bad and that their children need protection against this badness. We have reported the opinions parents had of neighbours' behaviour, their children's delinquencies or sexual precociousness; and parental suspiciousness of strangers believed to be sexual perverts. Not many of the chaperoning mothers were able to see life around them in perspective, their anxieties and their criticisms were expressed in terms of total condemnation. How could it be otherwise? A young child had been taken from a street play-group shortly before we interviewed and was found murdered. What the mothers felt and expressed in conversation with us indicated the shock and terror not just of that one incident, but of remembered events of loss of life at other times, in other places. These parents were driven into applying child-rearing measures which, under more normal conditions in a friendly and known neighbourhood, they would not be likely to apply. They kept their children indoors or under close supervision in the back yard; they accompanied them to and from school; they forbade them to play with undesirable youngsters in the streets. If the boys played out the mothers knew where to find them. These measures are applied at great cost to themselves. If child-centredness is defined as 'a willingness to take trouble for the attainment of aims in which the parent has no immediate interest' (Newson and Newson, 1968, p. 489), then, we believe, the exercise of chaperonage in deprived areas of the inner city is a child-centred form of parenting.

178 Chapter 11

PARENTAL METHODS AND CLASSROOM BEHAVIOUR

Following from all this an important question is whether there is any link between chaperonage, and its concomitant, strictness, and the boys' behaviour in the classroom. In discussing our findings the close association of chaperonage and non-delinquent behaviour, reported earlier in this chapter, must be borne in mind. Furthermore, delinquency of boys aged 13 was found to be associated with adverse behaviour ratings in the classroom three years earlier (see chapter 6). These results would lead to the prediction that absence of chaperonage should also be associated with poor adjustment in the classroom as indicated by adverse behaviour ratings.

The prediction is not borne out for the younger age group, but it does appear to be the case for the boys aged 10-11. As shown in Table 11.9 there is a consistent tendency for chaperoned boys to be 'better adjusted' than the unchaperoned boys on the sub-scales of our behaviour-rating instrument taken singly, although numbers are small and the significance criterion is not achieved. However, if we use a wider criterion of 'maladjustment' significant results do emerge.

TABLE 11.9 Chaperonage and classroom behaviour

Chaperonage:	6-7 years				10-11 years				
	Much	Some	None		Much	Some	None		
Behaviour:									
B(I)									
normal	7	9	6	NS	6	12	5	NS	
withdrawn*	2	2	2		1	2	3		
B(E)									
normal	8	9	6	NS	7	11	6	NS	
aggressive*	1	2	2		0	3	2		
A									
normal	8	7	7	NS	7	12	5	NS	
anxious*	1	4	1		0	2	3		
B(I), B(E), and A									
normal	6	5	4	NS	6	9	2		Significant
maladjusted**	3	6	4		1	5	6		at 0.05 level

* A score over 1½ standard deviations on the adverse side of the mean of the pooled control samples.
** An adverse score on any *one* of the above three subscales.

The discrepant results for the two age groups could indicate that the significant link between chaperonage and adjustment at 10-11 is spurious. However, the fact that the same trend runs through all the behaviour subscales indicates the possibility of a real relationship. Perhaps at the older age the outward effects of parental methods begin to show, which may be because at this age boys with little guidance or structure in their lives are becoming more prone to signs of stress or antagonism in facing the demands of school. Likewise, in escaping all parental surveillance they are at risk of delinquent influences from the neighbourhood. If this is a correct line of argument, then delinquency is not a product of personality disturbance or maladjustment, but both delinquency and poor adjustment are the results of lax methods of parenting. The findings merit further research.

Chapter 12

THE POVERTY SYNDROME AND CHILD DEVELOPMENT

The study described in this book was designed to explore the milieu of poverty and the effect this has on child development. The main sample was drawn from families known to a social services department, and the criteria for selection included residence in deprived areas of the inner city and having five or more children. Our assumption that these children would be severely socially handicapped in terms of specific environmental features known to be related to under-functioning or malfunctioning in school was verified in a subsequent investigation. Two controls were chosen for each focus boy from the main sample, nearest in age and attending the same schools but belonging to families not as severely socially handicapped as the main sample. An investigation of social handicap scores from a large, random sample of boys attending the schools used in the study indicated that the distribution of social handicap in deprived city areas constitutes a gradient from a sizeable contingent of families who are not socially handicapped to families who are moderately or severely socially handicapped. This finding confirms the findings of Barnes and Lucas (1975) discussed in chapter one; the majority of children in social priority areas are not members of very disadvantaged families. Nevertheless, our investigation of attainments in school showing considerable underachievement suggests the existence of institutional problems which are not wholly reducible to the problems of individual children.

Although the sample was selected from families known to a social services department, there is evidence, discussed in chapter 1, that in terms of income, housing, and size of family the sample were a subset of a population far in excess of the numbers in contact with the social services. In terms of personal attributes of the parents they should not be considered a special population with unique features. They found themselves in contact with the social services because of a crisis involving the wellbeing of their children. As discussed in chapter 2, half the sample had short contact with the department, and the other half were retained for preventive supervision. There were no salient differences between these two groups which we could ascertain. The decision to take no further action in some and not in other cases depends on a complex number of issues, among them the availability of a social worker in relation

to more pressing problems of other clients in the context of general
staff shortages; the family's ability to cope after the immediate
crisis had been solved by financial grant or support from relatives;
or because statutory obligations did not exist and the client was
no longer interested in obtaining help.

THE POVERTY SYNDROME

Families who seek help, or who are referred to a social services
department, have been exposed to stress over sometimes long periods
of time and this may result in behaviour patterns which may appear
to be 'deviant' or 'maladjusted', especially if the deviance is
expressed in motivational behaviour resulting in lack of ambition,
'shiftlessness', or 'irresponsibility'.

The kind of sociological model which is used as the basis of
analysis by the social work profession in accounting for observed
deviance is the following: the deviant person, or the deviant
family, is seen as behaving in a manner other than the ordinary,
normal, or average person or family, and consequently societal order
and stability are in danger of disruption. This analysis is usually
related to or seen in the context of the Parsonian explanation of
society as a functional entity with a shared normative order, but
as Becker (1963) has pointed out, it can equally be applied in a
conflict-model of society. Deviance is then explained in terms of
failure to obey distinctive rules considered important in mainstream
society, but not necessarily adopted by sub-cultures. Whatever the
basic assumptions about the nature of society and its normative
order (and sociological theories do not normally form part of a
social worker's training) the social worker who is confronted with
an assessment of the client's situation will take into account
deviant behaviour patterns. How important he judges such attitudes
to be in relation to the circumstances which caused the contact
between him and clients depends on individual cases and on the basic
approach of individual case workers.

Our approach is a different one, favouring an explanation which
sees the failure of the family to act in a protective capacity in
situational terms. Poverty groups are distinguished by their lack
of resources. Their chances of acquiring a secure income and for-
mation of assets are minimal, and the chances for socially upward
mobility are small because they lack educational opportunities. In
terms of class they are at the lowest end of the social hierarchy;
and in terms of social status they are held in low esteem by those
outside their own groups. Their position is characterized by total
powerlessness. The absence of skills weakens bargaining power in
the labour market. Dependence on welfare benefits results in sub-
mission to bureaucratic decisions. Non-participation in decision-
making results in political powerlessness in all but one aspect, and
that is the vote.

Rapidly rising living standards since the last war have been
shared, to a limited extent, by the lowest income groups, but the
earnings of the lowest decile remain obstinately below two-thirds of
average male earnings in industry, and consequently life chances of
this sector remain poor in comparison with the more affluent sectors

of the manual working classes. In a study of engineering workers in the East Midlands, Goldthorpe, Lockwood, Bechhofer and Platt (1969) noticed a 'normative convergence' between more prosperous manual workers and non-manual groups:

> involving in the case of white-collar workers a shift away from their traditional individualism towards greater reliance on collective means of pursuing their economic objectives; and in the case of manual workers a shift away from the community-oriented form of social life towards recognition of the conjugal family and its fortunes as concerns of overriding importance (p. 163).

In contrast to the more affluent sector, the families subjected to long-term poverty have learnt to develop adaptive patterns of living which are seen by others as the result of failure to improve their circumstances.

Changes in relative position vis-à-vis other working-class groups are also noticeable in housing. The large-scale exodus of the more prosperous sectors from deprived areas of the inner cities into new estates and new towns and the inward movement of those with fewer skills and resources has been discussed in chapter one. The difficulties that low-income families have in obtaining housing, and especially those with many children, has been documented by Rex and Moore (1967), who found that a family's chances to obtain housing depended not only on income and occupational status but also on their access to 'housing classes', a category employed by many housing departments which relates to variations in quality of housing and also to spatial distribution. Residence in older property situated in the more deprived areas of the city is typical for the lowest socio-economic groups.

There can be no doubt that residence in old houses, which are often damp, ill-ventilated and difficult to keep clean, is detrimental to health. In a survey of a large number of children Colley and Reid showed a definite social-class gradient in the frequency of chronic cough, history of bronchitis, and also diseases of the ears and nose. In an account of this research in the Report of the Chief Medical Officer of the Department of Education and Science for the years 1969-70 (HMSO, 1972) it is stated:

> When the recorded level of local air pollution was related to the frequency of chest illness a consistently higher level was found only among children of semi-skilled and unskilled workers. The trends in childhood chest illness were similar to those relating to severe and disabling bronchitis among adults in the same area (pp. 6-7).

The risks to health are manifold. Butler and Bonham (1963) investigated the causes of perinatal death and highlighted the increased perinatal mortality risk which is associated with unskilled occupational status, being 30 per cent greater than average. Douglas (1968) drew attention to the fact that:

> social class differences persist, or may even increase, during a period when there is a general improvement in facilities. The infant death rate, for example, has declined in all social classes, but the relative difference between the mortality of infants in the various social classes is as great today as it was 50 years ago (p. xii).

Wedge and Prosser (1973) studied disadvantaged children of the

National Child Development cohort and stated that poor physical
development is associated with a series of environmental disadvantages, among them the type of housing, circumstances at birth,
family diet, and the quality of medical care.
   A clinical examination of the school boys of our main sample and
half their controls showed increased rates of defects and ailments
among the severely socially-handicapped group, a higher proportion
below the twenty-fifth percentile of national norms for height and
weight, a significantly greater incidence of visual impairment and
a tendency towards more problems of hearing. Twenty boys from the
main sample and ten controls needed glasses, but only one in each
group was wearing them (Brennan, 1972). We have given an account of
the state of health of their parents in chapter 7; although this was
based on their own statements and not on medical evidence we gained
the impression that the incidence of illness, sensory impairment,
and disability was probably much higher than in a general population.
   The association of disadvantaged status, ill-health and disability,
although not experienced by all individuals, and not experienced to
the same extent, shows the self-perpetuating relationship of environmental factors and personal attributes, each reinforcing the other.
For example, two men from our main sample had irregular work
records due to disk trouble leading to prolonged unemployment, exhaustion of unemployment benefit and deterioration of general health.
(One should contrast these experiences with a similar condition in
the case of a person working in a professional or administrative
position who would be entitled to paid leave during treatment.) Such
stress situations tend to have a depressive effect if prolonged and
if they result in further adverse experiences such as lack of success in finding employment, harsh treatment on application for a
means-tested benefit, eviction and the status of homelessness, or the
removal of children into care. Chronic stress of this kind may result in feelings of failure, total loss of self-respect, or even
paranoid feelings of persecution, and these states of mind in turn
may lead to loss of motivation, suicidal actions, or aggressiveness
and homicidal tendencies. When family failure eventually leads to
contact with the social services it is not surprising that in many
cases personality attributes are seen as the main 'causative' factor.

SUB-CULTURAL CONTINUITIES

The idea of a 'culture of poverty' comes from Oscar Lewis (1967),
though it has been widely popularized and applied by others. The
essence of this approach is an adaptation of the concept of culture
to certain kinds of modern poverty in stratified societies with
egalitarian ideologies. Focusing on disorganization and personality
pathology in the ways of living of the poor, Lewis tries to show that
poverty is perpetuated primarily by 'cultural' elements in the traditional anthropological sense. The culture of poverty, he maintains,
provides people with 'a design for living, a ready-made set of solu-
tions for human problems' (p. 58). As Valentine (1968) has pointed
out, the literature on so-called culture patterns of the poor lacks
sufficient documentation to be convincing. More serious *doubt arises*
about the misapplication of the concept 'culture' to enforced ways

of living which are not cherished or held up as models of the desired life by those subjected to them. Lewis himself stated that 'the poverty of culture is one of the crucial aspects of the culture of poverty' (p. 58), thus demolishing the basic assumptions of culture theory.

While we do not consider the culture-of-poverty approach helpful in understanding the poverty pattern, we do not favour a simple situational model. Exponents of this model tend to stress the economic position, taking into account opportunities, skills, and long-term aspects of poverty. The important explanatory variable is seen as the situational one, and behaviour patterns arising out of the situation and in part adapting to it which persist through socialization processes are overlooked. The approach we favour is best described as a third model, the adaptational model. This takes account of behaviour as an adaptation to particular situations of deprivation which are passed on to the next generation in terms of commonly-held attitudes, beliefs and behaviour patterns. We see these as sub-cultural continuities.

We have consistently traced such sub-cultural continuities in that part of our study which dealt with child-rearing methods. The families who formed our main sample are rooted in the working classes and, as far as we were able to ascertain, predominantly in its non-skilled sector. We have described their methods of parenting and have compared these with the findings of the Nottingham study which comprises all social classes. In general, our sample of families conformed to patterns typical for the working classes in Nottingham when there were class differences. In many instances, however, they accentuated these class differences. For instance, while middle-class children on the whole like playing alone (at least some of the time), a quarter of Nottingham working-class children, but over half of the young children in our sample, did not like playing alone. Similarly, while 70 per cent of class I and II mothers with three or more children participated wholeheartedly in their children's play, the percentage in class V was 44, and in our sample it was only 24. Again, a gradation was observed in fathers' participation in their children's activities. The accentuation of class differences is, in many cases, accounted for by the constraints of poverty imposed on patterns of activity, and this is particularly apposite when it comes to comparing ownership of toys, activities and interests of the boys, and shared interests with their parents.

But there are two environmental factors associated with poverty which account for the accentuation of class-differences, and these are family size and quality of housing. In the Nottingham sample the majority of middle-class families were found to have two children or less, whereas the majority of the families of unskilled workers had four children or more; consequently methods of parenting reflect not just class differences in the abstract but specific characteristics. Where there are many children, individualized treatment is ruled out if the mother is the sole domestic worker. The quality of housing determines whether the family can afford to act autonomously within their own home, or whether proximity of neighbours and the absence of private play areas dictate methods in which consideration of neighbours takes priority over the interests of the children. Many of these child-rearing methods are traditional and adapted to typical family situations commonly experienced through generations. They become accepted methods of parenting in given circumstances, whether desired or not. The delegation of mothering to older siblings; early severance of mother-child contact and play in the unsupervised street play-group; encouragement of aggressive competitiveness; frequent use of repressive measures to curb undesirable behaviour instead of appeal to reason by language, all these methods and others were found to be practised among many working-class families in Nottingham and among almost all the families in our sample, all of whom had five or more children and lived in deprived areas of the city.

The impact that these patterns of parenting have on child development is severe. Early severance of mother-child contact and delegation of mothering to siblings affect the training of the young child in the achievement of age-appropriate behavioural standards. It does not stimulate language development. Children learn to adapt to situations of play in which only aggressiveness or withdrawal guarantee survival. The scarcity or total absence of toys and equipment suitable for play, and the absence of privacy allowing intensive play prevent the development of creative activities, powers of concentration, manipulative skills, and the re-enactment of experiences in imaginative role play. The absence of personal possessions deprives children of a culturally essential experience, the care of valued objects. Limited patterns of parental hobbies and interests and family activities deprive children of many enriching experiences, and thus many behavioural competences useful in school and in society at large cannot be developed.

It must be emphasized that sub-cultural continuities of this kind include a fair range of patterns, not all of which are represented by the entire sample, but the patterns we have here described in summary are sufficiently typical to merit description. However, a major finding of our study consisted of differences we encountered among the parents on a strictness-permissiveness dimension which proved to be closely associated with methods of supervision and consequent delinquent or non-delinquent behaviour. The practices described as 'chaperonage' are well-tried traditional ways of protecting children in an environment which is perceived as unsafe. In contrast, the families who did not restrict their children's activities seemed to adapt to adverse circumstances or seemed to perceive the environment in a way which did not demand any measures requiring a great deal of parental effort. It appears that the degree of strictness is a dimension of child rearing which is less dependent on circumstances than on parental attitudes and the interplay of temperaments of the two generations. In general terms both the stricter and the more permissive families had to adapt their management of the children to the constraints of their milieu. They shared many adverse experiences that limited the capabilities of their children.

We come to another and more profound level at which environmental constraints make an impact, and we refer here to the feelings of hopelessness and powerlessness which all families expressed at some time. Most of the families had a degree of awareness of their low status, and the low esteem in which they were held by others not in their group. There was an acute consciousness of the poverty trap and of the very limited opportunity structure in the market for jobs. Poor performance in the educational system was a parental memory reawakened by their children's experiences in school. Since upward mobility appeared obviously blocked, the only available solution to the pressures created by societal expectations was to develop a system of adaptive retrenchment. The objective is survival, the operative unit is the family. The needs of individuals must take second place. Decisions were made at family level and related to the main wage earner or recipient of benefit rather than to the needs of individual children. The value orientation of the child was thus shaped on a collective basis.

The effect that this may have on child development is suggested by Bruner (1971) who discusses the issue in a context of anthropological studies:

> It may be that a collective, rather than an individual, value orientation develops where the individual lacks power over the physical world. Lacking personal power he has no notion of personal importance. In terms of his cognitive categories he will be less likely to set himself apart from others and the physical world, he will be less self-conscious at the same time that he places less value upon himself (p. 30).

Bruner points out that in subsistence-cultures that require group action for survival, adults tend to evaluate a child's activities not *per se*, but in relation to the group, and a child's personal desires and intentions are discouraged. 'Thus a collective orientation does not arise as a by-product of individual powerlessness vis-à-vis the inanimate world but is systematically encouraged as socialization progresses' (p. 32). It is not clear to us how far the analogy can be taken in applying the insights gained in working with children from pre-literate communities to low-income families in a western democracy; but many practices observed by us appear to fit an explanation of this kind.

Children were generally seen by their parents in terms of the degree of usefulness or nuisance which they exhibited in family life. Punishment was usually administered indiscriminately to curb noise or conflict without regard for the wrong-doer. Money and other possessions were generally shared out irrespective of age-related needs. Chores around the house were allotted to whoever was available. Parents were reluctant to give individual children special attention. Members of the family adopted protective attitudes towards each other in facing demands from outsiders even if this involved statements or actions detrimental to the interests of individual children.

If there is a 'cycle of deprivation' involving generational aspects (most recently debated by Joseph, 1972, 1973a and 1973b) then the explanation lies probably in the survival strategy adopted by parents which overrides attention to individual needs of their children. Parental attitudes are rooted in deep-seated feelings of powerlessness. Their view of the world is dominated by mistrust and there is no curiosity about it. Thus the processes of socialization reflect and reinforce parental feelings of failure and perpetuate inequalities within society.

Chapter 12

THE BOYS IN THE CLASSROOM: ABILITY AND ATTAINMENTS

How did the boys present themselves in the classroom, and what are the specific problems that the teacher has to face in comparison with their less socially handicapped classmates? Our psychological examinations covered a wide area of cognitive functions of two age groups, one being in the upper part of the infant department, the other in the final years of the junior department of the primary school. It is not easy to draw together such manifold aspects into one picture, and one has to go beyond the immediate evidence to make a synthesis. In attempting this we include the control groups.

One firm outcome of our results is the impression of the wide-ranging differences on all types of measure. The spread of scores on most instruments is almost as wide among these inner-city boys as it would be in the national population. The mean is, of course, shifted towards the low end of the national average range on cognitive tests, and in reading attainments this shift is more marked, but the difference from the highest scores to the lowest is a big one. Not only is there this great range in any of the psychological characteristics, but our social-handicap instrument also shows the range in terms of some basic indices of disadvantage. The people of the inner city are much less a monotone than their surroundings.

In more detail, the control group of boys, who came from families scoring no, or only a mild, degree of social handicap, showed a mean score on a variety of verbal and non-verbal ability tests which is only a little below the national average (IQ close to 100). In contrast with this, they do much worse on reading: the 10-11 year olds have a mean reading age of about eighteen months below their average chronological age. Going from low social handicap through moderate to severe social handicap we found that there was a steep and significant gradation through the means of the three groups in ability test scores. With the low social-handicap group near one hundred, the moderate social-handicap group usually had means equivalent to an IO around ninety, and the boys from the main sample around eighty. There was only a slight tendency for the differences to be more marked in verbal tests. The exception was again reading, especially at 10-11 years, where the downward progression was not steep, and there was much more overlap between the three groups. At 6-7 years the two control groups had very similar means for reading, both poor, but the severely socially handicapped group was sufficiently worse to give an overall significant difference.

None of the boys in the main sample was very bright. In national terms the bulk of their ability test scores was around the eighty mark; in other words, it was below the national average band. Expressed in terms of the ability range within the neighbourhood the severely socially handicapped boys clustered around the lower half of the average group of their controls. These statements are, of course, generalizations. There was a fair range of scores on each of the ability tests; the standard deviation of our boys was not much less than that of a national sample. We picked out the consistently brightest quarter as judged by their mean scores on the major ability tests. There were thirteen boys in this group, four 6-7 year olds and nine 10-11 year olds. The best scorer in this group was just above the top of the average range for the control population, equivalent to a good average in national terms. The bottom scorer in this 'élite group' was just below the average score for the control population, about a low-average level on national norms. An important question is whether any strong relationship emerged between membership of this abler quarter of boys and any family patterns of child-rearing or othe family characteristics described earlier. In fact, no signs of any relationship could be established. The stresses that may be held to affect cognitive development in the entire group are shared by the abler boys among them.

The lack of significant inter-group differences on reading, especially at 10-11 years, and the poor showing of even the ablest boys at both ages, become a matter for concern. The existence of poor reading levels in the inner city is supported by the results of the EPA research (Halsey, 1972). This project has reported results for junior-school children (7-11 years) from four EPAs. The means for non-immigrant children in these four areas on the NFER Sentence-Reading Test 'A' are:

>   Deptford, London   93.0
>   Liverpool   91.6
>   West Riding   93.2
>   Birmingham   86.4

This test has a national mean of 100 and standard deviation 15. From the EPA data it might be inferred that the Midlands inner-city reading deficit might be due to lower general ability, since the Birmingham scores on the English Picture Vocabulary Test, with again a mean of 86.4, are also lower than those of the other areas. Our own data indicate that this explanation would be oversimple, because our low social-handicap groups were not low on ability tests, including the EPVT, yet remain behind on reading. The reasons are most probably multifactorial, and we suggest the following as arising from the Birmingham EPA data compared with the other disadvantaged areas in that research:

1 higher concentrations of immigrants, which may divert teaching effort;
2 higher rates of pupil turnover (probably due to slum clearance);
3 larger sizes of families in the inner city (see our discussion of this in chapter 3).

Probably similar results to ours and those of the Halsey study could be found in any inner-city zone with comparable problems. We must at this point ask why it should be that reading standards in such areas are so poor, when many local authorities do channel extra resources into the schools. We know this was the case in the thirty-nine schools of our sample, as it is in the Inner London Education Authority, the results of whose Literacy Survey have been reported in chapter one and will be discussed shortly. The search for root causes of poor reading achievement is extremely difficult. We would call attention to two basic points.

1 The school situation itself is rarely investigated as a variable. We believe more attention should be given to it, provided our expectations of schools are realistic. Most research on deprivation and education concerns itself with gross indices of community or individual disadvantage. We know from the work of Morris (1966) that children with poor reading are likely to be in the least well-equipped classrooms with the least experienced teachers.

Chapter 12

We know from the detailed study of Cane and Smithers (1971) that there is a big variation in the way reading is taught and that this has measurable effects on the attainment of children in schools with similar intakes. The evidence of multiple regression analyses by Peaker (1967; 1971) is that about 17 per cent of the variance in reading scores of primary school children can be attributed to teacher factors, which is not negligible albeit far less than the proportion accounted for by home factors. We would therefore consider that school should not be dismissed as an agent for improving reading standards, but also that the weight of other influences should be recognized, so that we do not expect schools to achieve the impossible. It should also be remembered that reading is one of the most teachable aspects of the curriculum, whereas other skills may be much more difficult to bring out.

2 There seems to be some independent effect from the characteristics of the entire neighbourhood in addition to the constituent individuals or their families, which always appear as the major source of differences in educational attainment. A study showing the effect of the schools' catchment area is that of Barnes and Lucas (1975, already referred to in chapter 1) which uses the ILEA's Literacy Survey of over 30,000 eight year olds. A multiple regression analysis of the reading scores showed a small but definite independent effect of the school's locality as measured by such variables as the proportion of children on free meals, from unskilled workers' homes, from large families, or who are immigrants, or who have poor attendance. On this evidence it is to be expected that within a given social class there will be a gradation of mean reading score in line with the relative deprivation of the school's catchment area. This is demonstrated by Little and Mabey (1973) using the same ILEA data, and is reported by us in chapter 1. The analysis of Barnes and Lucas shows that there still remains some effect of local community deprivation on the school itself. This independent effect is however small beside the influence of family (and presumably individual) characteristics. Barnes and Lucas estimate that, if the effect of the school's social context could be eliminated, the gap between the mean score of non-immigrant manual workers' children in the least and the most disadvantaged 10 per cent of schools would be reduced by 5.3 points (or months of reading age) which is substantial, though far smaller than the estimated reduction if the social factors affecting the child in the family were eliminated.

In summary then we would argue first, that the effects of strictly educational intervention in reading are often uninvestigated and underestimated, but also that they must be acknowledged as limited; second we would consider that the weight of problems within school catchment areas does to some extent depress attainment independently of family or child characteristics. Putting the two strands together, we can see why the extra resources devoted to the inner-city schools were not compensating for their problems, and why even the least disadvantaged boys were doing badly in reading. Our own reading data go beyond the ILEA evidence, since we included intelligence tests and administered all tests individually. The intelligence tests show the disturbing evidence of boys in the average range of ability achieving well below average in reading; indeed, this underachievement was also present in boys of below average ability. All this reflects badly on inner-city reading standards, whilst at the same time showing that the blame cannot be laid simply at the door of the school or the local authority. Educational action and the allocation of educational resources have a part to play, but we would join Barnes and Lucas in criticising the Plowden Council for emphasizing educational intervention when so much evidence points outside schools.

An obvious aspect of failure in our sample, apart from the reading results, is their poor showing on the various tests of verbal and non-verbal ability. What can we make of this? Some would argue that the results merely indicate that such tests are inappropriate for use on the disadvantaged even when, as in our sample, there is no problem of ethnic difference. We cannot ignore the fact that in some tests there will be a greater content of acquired experience (Cattell, 1963). Also, all tests require some basic motivation to do them which may vary according to social background. However, do these factors mean that human intelligence takes such different forms that we cannot use ability tests across the full range of our society? The same things which influence test scores must to some extent affect educational achievement and life chances, and therefore low test scores must reflect disadvantage in a real rather than illusory way. We acknowledge that low ability is not the total picture in social disadvantage. There is ample evidence that children are disadvantaged because their parents are (for example, Douglas, 1964; Douglas et al., 1968; Peaker, 1967, 1971; Jencks, 1972). However, this does not mean that tests give false results for the disadvantaged: it means that the disadvantaged are often doubly handicapped, both by their current ability level and the social group they were born into. Consistent low scoring on intelligence tests cannot be an illusory problem: such scales usually sample abilities over a long period. Of course, such learning is influenced by attitudes and motivations which are linked to cultural and economic factors outside the ability domain; but the all-embracing scope and longevity of the deficits indicates that they are real.

The teacher of disadvantaged children knows implicitly all we have said here: he sees children who seem below average on most intellectual skills; he sees that it is possible to rouse some of these to do excellent work, but often only in limited areas or for short intervals, when ingenuity and imaginative teaching tap some potent personal experience or some meaningful part of life; he sees others who are manifestly bright but fail to realize their potential. Many such teachers must reflect that the notion of 'true' ability never revealed seems for most children to be an increasingly unreal concept or, if real, to be rare and tragic. We are still left with the reality that the tests show, in part, the gap between what success in our society demands and what the children are equipped to give.

In summary, then, we do not naively believe in 'intelligence', nor in the tests as completely fair, but we do believe that the concept and the instruments have some validity and have something to tell us about the problems of the disadvantaged in their encounter with ways of thought and achievement that our culture not only values but finds important in practice.

Chapter 12

We now turn to another matter of controversy in accounting for our cognitive test results: the extent to which the inter-group differences are determined by genetic endowment. Many current authors have argued that in the Western culture of our time about 80 per cent of individual differences in intelligence are determined by genetic factors (Burt, 1958, 1966*; Jensen, 1969, 1972, 1973; Eysenck, 1971). Jensen in particular, but by no means alone, has gone on to infer that substantial parts of the differences between races or between social groups must also be inherited. Others have argued against all or part of Jensen's case, especially many authors replying to him in the 1969 Harvard Educational Review, and including Bodmer (1972) and Jencks (1972). The viewpoint taken will affect the approach to various forms of remediation.

The approach of Jencks is noteworthy in that he takes a different statistical approach to the same voluminous data that formed the basis of Jensen's work, yet he comes up with a far smaller genetic component: 45 per cent inheritance, 35 per cent environment, 20 per cent covariance. Covariance refers to the fact that bright parents usually have genetically advantaged children, and tend to rear them in stimulating ways; conversely dull parents tend to have disadvantaged children and to rear them in unstimulating ways. Environment is thus correlated with genetic factors and reinforces their effect. Covariance will also include the effect of a child's own intelligence on its environment. Jensen usually put covariance into the genetic side of the balance, since 'it is a result of the genotype's selective utilization of the environment' (1973, p. 54). Jencks, on the other hand, calls covariance 'a double advantage' phenomenon (1972, p. 69) and puts it on neither side of the scale. In addition to the covariance argument, Jencks also shows that heritability estimates vary according to the statistical method and type of sample used. One piece of evidence as to the latter is the higher heritability of intelligence in Britain compared with the USA, where the environment has wider extremes and thus creates greater differences.

Despite the scientific and theoretical importance of this controversy we must content ourselves with stating our point of view. First, to quote Jencks (1972, p. 76), 'Mathematical estimates of heritability tell us almost nothing about anything important'. In other words, the practical issues remain concerning educational intervention: Jencks is a useful corrective lest these get out of hand. Second, although it is implausible to assert that there is no genetic component whatever in differences between the average intelligence scores of social groups, it is unlikely that such differences are determined by genes to anything like the same degree as differences between individuals (Bodmer, 1972). Third, since the heritability of educational attainment is lower than that of intelligence and since it is the former which is most closely linked to occupational status (Jencks, 1972), we would be ill-advised to make too much of inheritance as a determinant of social differences. Fourth, all the reliable figures for heritability, environmental influence, and covariance are estimates of averages for whole populations and may well hide a higher environmental component or a higher covariance component among the socially disadvantaged because of their heavier burden of problems compared with the average in

---

* See, however, L.J. Kamin, 'The Science and Politics of IQ', Penguin Books, 1977.

the population. Jensen (1973) examines the evidence for this, but it has not yet been sufficiently investigated. For these reasons genetic influences seem to us an insufficient and unhelpful explanation. Even if there is a substantial genetic component, the remaining environmental component, including congenital factors (birth injuries, etc.), is large enough to make a substantial difference in determining a child's life chances. The practical and moral issues remain.

BEHAVIOUR RATING SCALES

We return to the findings of our study. Our results from the behaviour ratings, carried out by the boys' teachers, showed that there are more behaviour difficulties in the classroom with increasing social handicap. A Social Behaviour Rating Scale was constructed (Herbert, 1973, 1974b) that attempted to differentiate between anxious, withdrawn, and aggressive behaviour, and in addition, competent behaviour, meaning well-organized, persistent and resourceful behaviour. The 10-11 year old boys were also rated on a conduct scale used by West (1969). There are no national normative data on the Social Behaviour Rating Scale, and comparisons are therefore possible only within the main sample and the two control groups. We shall summarize the findings on this first. In the younger-age group the only significant, overall difference between the three groups is in terms of competence. The gradation down through the means is almost as steep as for the verbal ability tests; in addition, the younger boys from the main sample appear to be more withdrawn and inhibited than the least socially handicapped boys. For the older boys the competence scale again shows a linear gradation towards low competence among the severely socially handicapped boys, although not so sharply as among the younger boys. In addition, increasing social handicap is associated with increasing overt tension or worry, and an increase in dependent behaviour or behaviour demanding of adult attention. Finally, West's Conduct Scale shows much the biggest difference between the groups. This scale includes the following items: difficult to discipline, noticeably dirty and untidy, lazy, difficulties in relations with other children, truanting. It therefore identifies a mixed group of behaviour problems but also cognitive and competence elements. In line with West's findings a high score on this scale was significantly associated with increasing social handicap.

We would emphasize that the boys from the main sample did not, on average, show a predominant pattern of bad behaviour in their primary schools. This contrasted markedly with their high rates of delinquency three years later, but this was confined to a subgroup only. There was also a concentration of adverse behaviour ratings among this group, and we have, in chapter 11, traced the relationship of these behavioural characteristics and methods of child-rearing. Looking at behaviour patterns of the severely socially handicapped group as a whole it is obviously not possible to make the assumption that a boy who is not badly behaved in his primary school is less likely to get into trouble with the law later.

In a comparison of the three groups on severe adjustment problems expressed as a high score on anxiety, withdrawn behaviour, or aggressiveness, it is noteworthy that there are consistently almost twice as many boys from the main sample, indicating that the incidence of severe behaviour difficulties is higher among the very socially handicapped boys, although none of the comparisons are statistically quite significant. On the other hand, if we adopt the criterion of 'good adjustment', meaning a score no higher than the average for all three groups, there are significant differences between them: 21.4 per cent of the controls are well adjusted, but only 7 per cent of the main sample boys (significant at 5 per cent level).

The findings on the differences in competence deserve attention. As discussed in chapter 4, the competence scale has moderate to high correlations with reading, especially among the boys from the main sample. These probably arise from the fact that the competence scale measures what its name suggests: effective, self-organized, reliable, educationally motivated behaviour. In short, a teacher sees the competent child as knowing what being in school is about, not merely learning to read but getting on with the job, needing little supervision, organizing himself, and so on. The competent boy is the 'good pupil'. The child with severe social handicap is often not a good pupil in this sense.

The origins of competence are obscure. It is not merely a matter of being reasonably bright; there is, for example, a sex difference in favour of girls (Hague, 1970). Bernstein and Davies (1969) suggest that

> many working-class fathers do not support and develop the educational role of their children. It may be that, as a result, the working-class boy does not come to value his educational role as this is not identified with his most significant male relation.

Responsibility for the child's education in the primary school rests largely with the mother, and the situation may be reinforced by the 'almost totally feminine world of the infant school' (p. 67).

We would see competence in school as does Jencks (1973) in relation to job competence in adult life: 'The definition of competence varies greatly from one job to another, but it seems in most cases to depend more on personality than on technical skills' (p. 8). Because school is a continuing learning process, the job requirements are more cognitive than in adult employment settings, but to us competence is nevertheless an amalgam of traits which are the bridge between straight cognitive potential and the realization of that potential in ways which are required by society. It so happens that most children who are bright tend to be highly competent, both because their prowess makes it easy, and because they more often come from homes which prepare them for school. It is perfectly possible to find a very bright child who by reason of maladjustment, immaturity or sheer conconformity is not very competent in school. Similarly, although it is more difficult for intellectually average or below-average children to acquire high competence, they can show varying degrees. In school, competence is the business of knowing how to go about things and it is the characteristic which the child with severe social handicap is not geared to acquire.

One fairly clear conclusion of our findings is that the boys' teachers do not operate a cognitive or social 'halo effect', whereby positive or negative assessments in one respect lead to similar assessments in others. The halo effect as it operates in the classroom would lead one to expect that teachers would tend to rate poor boys or dull boys as difficult. The many teachers involved in rating the behaviour of 174 boys in fact clearly distinguished between social status, academic ability, and social conduct. The finding is a healthy corrective to the often expressed belief that teachers tend to be prejudiced by their perception of social disadvantage or relative brightness.

A further reflection on the behaviour ratings must be that we have been dealing with boys up to 11 years old. None of our results can really illuminate the problems of difficult anti-social conduct among inner-city children in their teens. There were indeed some extremely difficult boys in all three social-handicap groups at both ages; it would be a miracle if they did not continue to be severe problems in the secondary school. However, although the problems were doubtless more severe than in suburban schools, none of our primary schools was experiencing the degree of difficult behaviour, absenteeism, or apathy that would be apparent in secondary schools in the same neighbourhoods. Our guess is that our results depict the beginnings of the problem: the presence at 10-11 years of a substantial nucleus of extremely difficult boys, generally poor reading standards, and self-evaluations that are not based on academic achievement, as evidence by the low correlations of the self-concept scales with academic measures. For increased difficulties to emerge at secondary-school level, a substantial number of the boys who do not show behaviour difficulties at 11 years must begin to do so. A compelling question arises as to the development of the very quiet, anxious, or withdrawn boys - do they become sullenly defiant, victimized, or merely submerged in what one hopes is still a majority (if reduced) of pupils who conform in at least outward appearance with the requirements of school? Or are they the nucleus of chronic non-attenders? We do not know. Whatever happens, it seems that the germs of disaffection or indifference are visible in the primary schools.

THE SELF-REPORT SCALES

The 10-11 year old boys completed two self-report scales, the New Junior Maudsley Inventory (Furneaux and Gibson, 1966) and a self-concept questionnaire. In brief, the results indicated that there were no inter-group differences between the main research sample and their control groups in terms of the Extraversion, Neuroticism, or Lie (unrealistic self-approbation) subscales of the NJMI, nor in terms of the self-concept items, where the boys rated themselves on Evaluation (physical characteristics, school performance, peer-group acceptance, etc.) and Motivation (aspiration).

Closer examination of the self-report scale results indicated that the control boys seemed more able to put themselves consistently at one end or the other of a 'good-bad' continuum, that they agreed more often with their teachers' ratings on dimensions related to extraversion, and that their ratings of their own motivation was reflected in reading and competence scores (high motivation with high scores). Thus the control boys seemed to have a more consistent self-image, perhaps because they were more aware of the yardsticks by which society judges people.

WHAT CAN BE DONE?

New concepts of schooling

The growing problems of the schools in priority areas have led to the question whether the education system is too rigid and is failing to change so as to meet the needs of children in the inner city. The establishment of 'community schools' was one of the objectives of the EPA action programme (Halsey, 1972). It was hoped that the involvement of parents in their children's education could be stepped up and 'a sense of responsibility for their communities' among the people living in deprived areas could be increased. A proposition of this kind appears to assume a strong local social system in deprived areas; while this may be true in some, it is not characteristic of most. Whether increased parental involvement, given a stable community, will have a lasting effect cannot be foretold on the basis of EPA experience. Experimentation outside the local authority education system is described by Newell (1975): 'When offered a radical alternative, children are the quickest to see that the conventional state school was not designed with their interests at heart'. The White Lion School, like other free schools of this type, attempts to create a new institutional model 'free as far as possible from the constraints of mass schooling, flexible and aimed at helping individuals *of all ages* to identify and meet their learning needs'. It may be that small, informal, educational resource centres of this kind, which are accessible to children and to their parents, will prove to be more effective in stimulating learning motivation and in achieving contact between parents and teachers which is meaningful for both parties. The approach is very different from that described in the EPA study in that it invites the local community to say what should be done, but doubts have been expressed whether community control is compatible with a professionally organized local authority school (Musgrove and Taylor, 1969).

Increased parental involvement does not necessarily result in harmonious relations with school staff or improved functioning of children. Halsey, well aware of the potential culture clash in the educational priority area, points out that 'ideas, values and relationships within the school may conflict with those of the home, and that the world assumed by teachers and school books may be unreal to the children' (ibid., p. 43). In deciding what values and attitudes are to be perpetuated or to be regarded as undesirable the teacher will have to make moral judgments 'about the ways of life that the schools should support and those they should try to change' (ibid., p. 118). If this is to be part of the teacher's role one may well ask, are teachers in training being prepared for such judgments, and if they are, what are the moral judgments to be? The Plowden Council was deliberately vague (Peters, 1969).

Intervention programmes in nursery and infants school

The evidence on the efficacy of nursery education for the disadvantaged is not encouraging (B. Tizard, 1974; Smith and James, 1975); certain programmes, especially if structured to achieve cognitive goals, produce gains, but these seem to 'wash out' in a year or so, even if there is involvement of the home through a visiting service. However, Jack Tizard (1975) puts forward a set of points to justify the growing demand for pre-school services which may be summarized thus:

1 Nursery education shows 'intrinsic rather than instrumental concern with the happiness and well-being of young children. In providing for these, nursery schooling has been conspicuously successful' (p. 217).

2 Although there is 'no magic' in the early years of development, and therefore long-term results are not to be expected, nevertheless specific skills and behaviours can be helped in the pre-school in such a way as to prepare for initial coping with school. 'Further analysis of the competencies required by five year old children to be ready for infant school is needed before we can decide upon the ways in which we can best prepare them for this and examine the success of our efforts. The problems seem by no means insuperable' (ibid., p. 219).

3 Benefits to mothers as well as children may be a major aspect - the freedom to go to work and to be away from the constant pressures of their children. Also, the nursery can act as a social centre and a source of parent education.

Tizard's arguments carry special force in the case of the severely disadvantaged family which is dominated by stress; interpersonal, economic, and physical. The children may be bewildered or miserable and may need a place where their 'happiness and well being' are the prime concern. A small proportion of such families may place their children in danger of neglect or non-accidental injury; these children are priority cases for nursery places. The deficiencies which readily lend themselves to analysis to provide specific programmes are deficiencies in verbal skills and the ability to cope with the daily routine in the nursery and to master the environment. These are skills which depend largely on the self-confidence derived from the process of trying, correction, encouragement and success, backed up by the aid of models of competence. Many of these goals are part of the 'competence' factor in which our own results have shown disadvantaged children to be in need of help. The aims of nursery education for the disadvantaged child need to be specific. A good way to achieve this might be for the staff of primary schools and local pre-school agencies to work out together what these aims are, to ensure both that the pre-school is geared to particular ends, and that the primary school continues the work in a setting and manner which are not strange to the child and can be understood by his parents.

In a period of low economic growth the priority should be on the creation of pre-school provisions for disadvantaged children, and this might be achieved by co-operation of the social services and education departments. In addition, there should be an attempt to concentrate existing resources of primary schools in deprived

areas to achieve the following:
1 a sense of purpose and mastery in the children;
2 the social skills needed to become a member of the school community.

Both (1) and (2) are once more part of the competence skills in which disadvantaged children are deficient.

3 the language and comprehension required in explaining and controlling the world, and, at a later stage, in acquiring literacy and numeracy.

Many primary schools already address themselves to these aims, but probably much more could be achieved through inservice training programmes and work on resource materials and curriculum development to add to the efficacy of such extra teachers and assistants as can be provided.

Intervention at later stages of schooling

Whether or not radical new forms of schooling are set up, the great majority of children will still have to be educated in local authority schools using existing buildings and staff. In many ways our views on the objectives of these schools can be stated in the same terms as for pre-school education: there is the same need for individual care, experience and encouragement of mastery, and structured achievement in language, basic literacy and numeracy, together with co-ordination of approaches at the transition from one age level of schooling to the next. More specifically, we feel the following points need stressing:

1 Because of the high incidence of learning problems (and often behaviour difficulties) in priority areas, it seems superfluous to use screening techniques to 'identify' children. It is nevertheless important to train teachers in sequential diagnosis and prescription.
2 The all-round poor performance of the disadvantaged, as illustrated by our results, should lead to all-round programmes, but *not* to the assumption that non-verbal and physical skills should be trained before literacy work is started.
3 'Remedial' work should not be regarded as something which goes on for part of the day, but rather as integral with the whole curriculum; in other words, the remedial teacher should be an adviser to the whole staff on methods, materials and language content.
4 Because some children will be spasmodic attenders, the curriculum should be geared to sequences for individuals rather than groups.
5 The implications of the 'chaperonage' concept should be explored in the context of school. Children need to be watched and cared about, and chaperonage should be taken beyond mere negative restrictions to become a means of creating a base from which autonomy and competence can grow.
6 Our evidence that the absence of chaperonage at home is linked with delinquency and poor adjustment in the classroom indicates the need for the school to make informal and helpful contact with families showing this pattern before trouble starts.
7 Severely disadvantaged boys appear to have an increased preva-

lence of anxious or dependent behaviour in late primary school. We need to ask whether the school itself is partly responsible for this, and to take early steps to overcome it.
8 School attendance difficulties in the disadvantaged child should be recognized as a deep and continuous problem, which cannot be solved by any of the usual combinations of persuasion, pressure, and care proceedings. We need more flexible resources and placements, including some schools or units with highly individual approaches. The concept of residential 'assessment' with its implications that there are personal problems which can be assessed away from home, should be dropped for such children, and the residential facilities should be used for creative and positive purposes.

HOPE

We have enumerated some educational measures which we believe to be beneficial to disadvantaged children. We have done so in awareness of all the evidence which now exists that the influence of nursery and school cannot counteract the effect that poverty has on child development. Educational action has a part to play in making good the damage which is done; it may help the child to adapt to the demands of school and to benefit more fully from the range of social activities that schools have to offer. But compensatory measures in the nursery and at school do not alter the long-term educational prospects of disadvantaged children. Our views are strengthened by the findings of Jencks (1972) who, in a comprehensive, critical analysis of the American problem of social disadvantage, states that schools are unqualified to make the changes that would bring about equality of opportunity as long as the absolute level of inequality persists in society. The same is true in Britain: the promise contained in the 1944 Education Act of equality of opportunity for all children has not yet been fulfilled.

Equality of opportunity is one of the basic principles of the democratic state. To make this a reality it is essential to eliminate poverty. Poverty is not generated by 'inadequate' people who conglomerate in deprived areas; it arises in a social system in which low wages, inadequate welfare provisions, a chronic shortage of housing, and unemployment are allowed to exist. Thus, in the final analysis the problem of disadvantaged children does not lie in genetic or in psychological deficits, it lies in an unequal distribution of the resources of our society. The position of the children who grow up in poverty is one of hope, because their disadvantage, given the will, can be eliminated.

APPENDICES

Appendix A
JOHN VEIT WILSON

Note: This appendix refers to the survey of the sources, level and duration of the families' income which has been described in chapter 2. The first section describes the sources and limitations of the data and how they were used; the second elaborates the findings on the payments of social security benefits.

1 METHOD: SOURCES AND LIMITATIONS OF THE DATA

Sources and duration of income

It has been assumed that, during the periods not covered by social security payments, the families were dependent upon the men's earnings. This does not mean that the men were necessarily in work during the periods in which they were not drawing social security benefits. In 1972, for example, one-fifth of all men and women nationally registered as unemployed were receiving neither unemployment nor supplementary benefit (Field 1975). Payment of social security benefits is conditional on not being in fulltime work. Our estimates will thus tend to exaggerate the periods in work, since they are based only on claims for benefit. Even if the cynical assumption is made that some claims were concurrent with undisclosed employment or self-employment (about which we have no evidence from the sample), the Report of the Committee on Abuse of Social Security Benefits (1973) showed this to be miniscule by comparison with the number of claims made. On its statistics we should not expect to find that more than one or two of the many hundreds of claims for benefit during spells of sickness or unemployment made by the men in our sample were fraudulent.

Whether or not the fathers were working, the families in the sample must have been living on current or previous earnings if they were not claiming social security benefits. Consequently, we speak of 'dependence' on such earnings during periods when there was no dependence on social security as the prime source of household income.

Income from earnings

The fathers were asked to say what their earnings were at the time when they were interviewed, during 1969-70. Where the men were not available, the mothers were asked instead. The answers on 'take home pay' which we were given are, therefore, subject to the unreliability generally attached to this sort of response. On the other hand, there is no reason to believe they were any more unreliable than the data on earnings collected in the same way by many other household surveys. On the contrary, the ongoing relationship between the research staff and the sample families may have encouraged a greater degree of trust and accurate reporting than may be achieved by the impersonal single doorstep interview.

We attempted to complement or corroborate the earnings data with later information provided by the Department of Employment or the DHSS. The P60 statements of previous annual earnings were useless for our purpose since they were annual totals and could not even give averages for weeks in work. The SBC data were gross earnings, but often more up to date than the men's reports. To get a picture of the mid-1970 earnings, we took the SBC data where they existed, and made a standard £1.00 deduction from it to allow for National Insurance and possible tax deductions. This deduction is on the low side and thus overestimates net earnings. Where there were no SBC data, we took the men's own reports. Where they declared a range of earnings (for example, £20 to £22) we took the mid point (£21).

The earnings data should, therefore, be treated with caution. In taking earlier earnings and comparing them with a 1970 poverty line (see below) we may be overestimating the extent of poverty among the men in work, and this effect may be exaggerated by any tendency to underreport earnings received. On the other hand, if the men exaggerated their earning power, and if they had more deducted from their gross earnings than we have allowed for, then this effect is reduced.

Income from social security sources

The DHSS supplied us, as far as possible, with the following information:

Supplementary Benefit

(a) dates during which claimed in 1969-70
(b) rate(s) of payment
(c) rent allowance
(d) discretionary extra circumstances allowance
(e) exceptional needs payments - number, amount and reasons
(f) details of wage stop imposed, if any
(g) information about earnings in between claims for supplementary benefit
(h) details of wife's earnings, if any
(i) details of non-dependents in the household

Appendix A

Contributory Benefits

(a) dates and details of claims
(b) rates of benefit showing earnings-related supplement separately
(c) information (from P60 forms) on earnings in 1968-69 and 1969-70

In addition, they were able to give us detailed breakdown of the way in which the rates of payment were calculated for twenty-six of the families in mid-1970. These detailed forms provided information on the ages of the dependent children, and on such adjustments to the scale rates of payment as deductions for voluntary savings, for proportions of the rent assumed to be paid by non-dependent members of the household, for rent arrears or rent paid direct to the local authority housing department, and for the wage stop. It also gave information on additions to the scale rate such as the long-term addition (these terms are described below). It noted income from other sources received by the claimant and taken into account or disregarded by the SBC in calculating the claimant's entitlement. This information was exceptionally useful to us in calculating the cash income of the families receiving supplementary benefit, and we treated it in the following ways.

Where a family had 'voluntary savings' deducted from its supplementary benefit, we did not treat the deduction as income since it was not available as such to the family at that time. Instead, we totalled all the repayments of 'voluntary savings' and all the exceptional needs payments paid to the family during the twelve month period up to the end of the month in which fell the week's income we were analysing. We divided this total by fifty-two and added this sum back as weekly income. In a handful of cases the dates of the payments were not given, and the two years of payments were divided by 104 weeks instead. However, where possible we preferred to consider the contribution to the level of living experienced by the family as that provided by payments made before the week in question; we felt that subsequent repayments, which may have been made in different circumstances, would have confused rather than clarified this picture.

Rent paid direct to the landlord (the local authority housing department) was included as income, since it entered into the assessment of the poverty line as well (see below). But proportions of rent assumed to be paid by non-dependent members of the household (for example, children who had left school and started work) were added back as income at the same rate as the SBC had deducted them from their payment, since, if the non-dependent member had not been there, the SBC would have paid the full amount, if it were reasonable in the circumstances (see below). The method of calculating rents is described later. On the other hand, payments made by the SBC direct to the local housing authority for the rent arrears of tenants were treated as deductions from income, since they were analogous to voluntary savings, i.e., had they not been made as regular deductions, the SBC would have been empowered to make them as additional exceptional needs payments.

The scale rates are the various cash sums laid down by the SBC as the minimum amount required by claimants of various statuses and ages. The relevant sections of the scales in mid-1970 were:

Married couple, householders	£7.85 per week
Single person, householder	£4.80 per week
Dependent children and non-householders aged	
0-4	£1.40 per week
5-10	£1.65 per week
11-12	£2.05 per week
13-15	£2.20 per week
16-17	£2.80 per week
18-20	£3.20 per week
21 upwards	£3.85 per week

The scale also includes the actual housing cost of the claimant.

Voluntary savings under supplementary benefit is an arrangement whereby the SBC makes deductions from the full scale payable to regular recipients, so that the claimant family has a reserve of money available to cover irregular expenses. Interest is not paid on such savings, nor are claimants provided as of right with a savings book or other official record of the amount standing to their name. They can, of course, enquire, and can make withdrawals up to the amount saved. If expenses occur which exceed the balance saved, a claim for an exceptional-needs payment would have to be made over and above the scale rates.

The wage stop: 'In cases where the payment of a supplementary allowance is made conditional upon registration for employment, the Act requires the Commission, unless there are exceptional circumstances, to limit the allowance so that the claimant's income while unemployed is not greater than it would be if he were engaged in full-time work in his normal occupation.' Supplementary Benefits Handbook, HMSO 1971, para. 68.

The long-term addition: A standard sum (of 50p. at the time) added to the payments to 'people who have received a supplementary allowance for a continuous period of two years and whose allowance is not, and was not at any time during that period, subject to the condition of registration for employment.' Ibid., para 51.

Rent is reasonable in the circumstances if the rent is reasonable for the accommodation and the accommodation is reasonable for the tenants. Ibid., para. 44.

The poverty line

Supplementary benefit scales and augmentation

We have used the same measures of poverty as were used by the DHSS Statistical Research Division in their study of families with both parents present (Howe, 1971). These measures took the scale rate plus the appropriate housing cost (see below) and augmented them by a sum equivalent to the average weekly exceptional-needs payments received nationally by long-term unemployed recipients of supplementary benefit. These sums were approximately 5p. for a family containing one or two children, and 5p. for each extra child. By including this augmentation, the poverty line represents more closely the average cash income received by those families certified as being in poverty by their receipts of supplementary benefit. It enabled us to discover how the additional payments made to the families in our sample compared with national experience.

Families whose income was derived from employment were in a slightly different position, since a part of the net wages received would have to be devoted to expenses connected with the employment, such as travel to work, union dues and special clothing. We therefore used the same additional augmentation of the poverty line as the DHSS study to give a basis for comparing incomes in work, that is, for 1970, a standard increase of 50p. in the poverty line. It is arguable that we would have achieved a similar effect (with slightly different percentage results) by reducing the net earnings. But we felt it proper to show the 'basic needs' of the family as greater by this amount when the father was in work.

In short, the poverty lines we have used are, for families on social security, the appropriate scale rates including rent plus the exceptional needs payments augmentation, and for families in work, this sum plus 50p. for contingent employment expenses. Where we have compared incomes in and out of work, these are based on the two different poverty lines appropriate to the same family.

Family composition

In calculating the appropriate scales, we have not included non-dependent members of the household, such as children who have left school and entered employment. We did so because we could not accurately ascertain whether or not they were present or what their incomes were. We made the assumption that any contribution which they might make towards rent, food and other goods and services they received, would equal the cost of these goods and services and would therefore not be available as income to the rest of the family. Children technically dependent who were in effect permanently absent from home, for example, because they lived with relations or in residential schools, were also excluded. In such cases the family allowance is generally payable to the person with charge of the child, either directly or by the parent. The information from the DHSS made it clear that this happened in a few cases. Where such an absent child is excluded from our calculation of the scale rates, the family allowance received by the parents has, therefore, also been excluded.

The DHSS data showed that some families were receiving social security benefits calculated on the basis that the father was absent from home. In such instances, we calculated the poverty line on the same basis, that is, for a single-headed household. However, in a cohabitation case, where a supplementary allowance was paid for the children alone, we have calculated the scales on the basis of a complete family and have added a notional sum equivalent to the scale rate for a married couple onto the income received from social security sources, since we have assumed that the father made at least such a contribution, and had he not been at work the DHSS would have paid this sum.

Housing costs

The normal SBC practice is to pay claimants the amount of their rent if it is reasonable in the circumstances, or the interest, but not the capital element, of mortgage repayments to owner occupiers with loans to repay. Rates are also payable in full by the SBC. These sums must, therefore, be included in each family's poverty line 'needs'. Some local authorities, including the landlord of all but four families in the sample, do not charge rent every week of the year but allow a certain number of 'rent free' weeks. This landlord allowed four such weeks, so that rent was only payable for forty-eight weeks in fifty-two. In such cases, the SBC do not pay the full weekly amount payable by the tenant, but one fifty-second of the annual total. In assessing the housing cost to include in each family's poverty line, we have, therefore, taken 48/52 of the amount declared as rent paid by council tenants. We found a high degree of correspondence between our calculations and the amounts actually paid by the SBC, variations rarely amounting to more than a few pence. In the few cases of variation, we took whichever figure was the higher, the tenant's or the SBC's. The effect of this arrangement is that the cash housing cost received from the SBC by such local authority claimants is slightly less than the amount they have to pay their landlord. They are thus slightly worse off for forty-eight weeks of the year, but receive the same rent allowance, although they do not need to pay it, for the remaining four weeks.

Three of the families were the tenants of private landlords and had to pay rent every week of the year. Their rent was therefore included in full. One family was buying a house on mortgage. We were not able to ascertain the payments made. But since they were deducted from the father's earnings at source, we have calculated this family's poverty line on the basis of the omission of housing cost from both poverty line and income, producing a similar though not identical result.

2 INCOME FROM SOCIAL SECURITY SOURCES

In chapter 2 we have stated that the majority of families had more deducted than added to their scale incomes from supplementary benefit. In this section we give the detailed findings which led us to this conclusion.

Additions to social security payments

While the fathers were out of work and the families were dependent upon social security incomes, the total income they received could have been greater than the scale rates of supplementary benefit if they had received additional payments from social security sources or had income from other sources which was disregarded by the SBC. Among the twenty-five families for whom we have detailed supplementary benefit calculations provided by the SBC, only one had disregarded income (£2.00 disregarded of an industrial disablement pension of £3.35 a week). Among the eight families for whom we did not have a written explanation of their allowances, only three had incomes above what we calculated as their entitlement. One of these families appeared to receive £2.26 a week over the poverty line; the other two only 21p. and 10p. Since none of the thirty-three families were receiving national insurance benefits without supplementary benefit, we have no reason to believe that any of them were receiving earnings-related supplement during the period of our calculations at a level high enough to make them ineligible for supplementary benefit or, in other words, above their poverty line. The family with the £2.26 above the poverty line may not have been receiving so much if we have underestimated the household composition and age. The father had been paid unemployment benefit plus earnings-related supplement in the past, but his entitlement had been exhausted by the middle of 1970. Thus the only source of additional income beyond the scale entitlement for the thirty-three families was the SBC, and it had provided extra income to just over two-thirds of them (twenty-four) currently, or at some time in the previous year, while they were dependent upon social security sources

Among regular weekly allowances, the long-term addition was paid to only just over half of the nine families continuously dependent on social security. They may have been unemployed or had not been claimants long enough (two unbroken years) to qualify. The remaining twenty-four families in Groups 2 and 3 did not, of course, qualify for this addition. Only one family received a regular discretionary allowance, in its case for extra fuel at the rate of 17p. a week. We did not have, nor had we asked the DHSS for, sufficient information to tell why this family and not others in similar situations received this allowance, or the reason for the amount. There were no other regular allowances.

Irregular payments to cover exceptional needs (for example, for consumer durables and work tools and clothing) were more common but varied widely in their incidence. Moreover, both the incidence and the average amount seemed to vary in relation to the duration of dependence on social security. This may be because the irregularly dependent families in Groups 2 and 3 tended to replace household goods during their periods of employment and thus made fewer claims. Or it may reflect the differential use of their discretionary powers by the officers of the SBC in paying, modifying or refusing claims. In the absence of qualitative information, we cannot hazard an explanation. What is clear is that all the families in Group 1 and nine out of the ten families in Group 2 had received exceptional needs payments during the previous year, but that only six of the fourteen families in Group 3 had done so. Moreover, the average weekly value of these payments ranged from 60p. among Group 1 families, down to 28p. among the six families in Group 3. Even though nearly all the ten families in Group 2 had received one or more payments, the weekly value of them was closer to that of the Group 3 families than to those continuously dependent on social security.

The cash amounts paid by the SBC of course varied considerably according to the items to be bought, but (judging by the information given) they can hardly be considered as unduly extravagant. For example, one family was paid £3.75 with which to buy 'a bed and mattress'; another was given £19.95 'to cover the cost of three pairs of double-sized blankets, two pairs of single-sized blankets and one man's suit'. Given the range of prices known to us even for poor quality goods, we find it hard to understand how the officers of the SBC calculated these sums, unless they were for second-hand goods. If this were the case, it would not be in conformity with the policy of the Commission itself. (Personal communication to the Child Poverty Action Group from Lord Collison, Chairman of the SBC. The practice was not unknown elsewhere and was by no means confined to the West Midlands Region.)

One family (in Group 1) appeared to have benefited by a calculating error on the part of the SBC officers. A daughter aged eighteen lived at home but was unemployed and received supplementary benefit in her own right. She was paid the standard rent allowance of 55p. (for a claimant living in someone else's household). But the family's own SB rent allowance was reduced by only 17p. (the proportionate amount for a non-dependent in the household). The family was thus inadvertently 38p. better off.

Additions to the scale social security incomes of these families were thus small, and made little difference to the total amount - although any sum at all was valuable to families living on such low incomes for so long a time.

Deductions from social security payments

A majority of the thirty-three families had incomes which fell below their appropriate scale entitlements. This was either because the SBC made certain deductions before paying them, or because of errors and assumptions in their calculations or ours, and because of our inclusion in their poverty lines of the average exceptional needs payments (see above). One family was paid £1.00 too little by the SBC because its officers had over-estimated the number of family allowances actually received (and consequently deducted from the cash entitlement to supplementary benefit). We had inadequate information on eight other families, but where we were able to make comparisons between our methods of calculation and the facts available to the SBC we have only found small differences. We therefore have no reason to believe that this detracts significantly from the findings presented here.

Four kinds of deduction by the SBC were found. (a) Seven families in Groups 1 and 2, those continuously or lengthily dependent on social security, had sums ranging from 50p. to £1.50 deducted weekly from their allowances as 'voluntary savings' (for definition see section 1). (b) Five families had sums for rent arrears deducted and paid direct to the local authority landlord. These sums ranged from 13p. to 50p. weekly. (c) Seven families had their supplementary benefit wage stopped in mid-1970. These were in Groups 2 and 3, intermittently dependent on social security. A larger number of men had been wage stopped at other times. Only two men of the fourteen in Group 3 were wage stopped; but among the ten men in Group 2 with a longer duration of dependency on social security, four were wage stopped. One man was stopped to £13.12, and the remaining five to between £16.06 and £16.60. SBC procedure took local authority basic wage rates for labourers (set by the National Joint Council) as a guide to the level of income to which to wage stop. In mid-1970 these were about £14. But these levels should only have been used where evidence of the previous or potential earnings of the men is not available. The two men in Group 3 for whom we had earnings data were in fact wage stopped to levels within £1.00 of their declared earnings. (d) One man was stopped the standard 75p. because he had left his work for reasons considered unacceptable; he was categorized as 'voluntarily unemployed' and his family penalized for it.

Not only did these official deductions reduce the families' incomes, but they were aggregative. Two families had both voluntary

207 Appendix A

savings and rent arrears deducted, totalling £1.75 and £2.00. Another family, already wage stopped by £3.06, had rent arrears of 50p. deducted (but the effect of this was counteracted by the only rent rebate we recorded, payable solely because of the wage stop).

The net effect of additions and deductions

The table below shows that the majority of families had more deducted than added to their scale incomes of supplementary benefit. Among the reasons for this was that none of the seven families with 'voluntary savings' deductions recovered, over a period of a year, as much from the SBC in withdrawals or ENP's as they had 'paid in'. They might have been better off if they had not attempted to save.

The families continuously dependent on social security sources (Group 1) received on average slightly more in ENP's than did supplementary benefit recipients nationally. Some of them also received long-term additions. But these additions were outweighed by the greater average 'voluntary savings' deduction. The families in Groups 2 and 3 were on average worse off because of the effects of the wage stop and, in Group 3, because they received fewer ENP's than did the average family nationally. The latter effect was small.

TABLE A.1 Additions to and deductions from supplementary benefit scale payments of groups 1-3

Group	1 Continuous soc. sec.		2 Mainly off work		3 Intermediate	
Number of families	9		10		14	
	No. of families	Amount weekly pence	No. of families	Amount weekly pence	No. of families	Amount weekly pence
ADDITIONS TO SB SCALES:						
Long term addition	5	50	0		0	
Discretionary allowances	0		1	Fuel 17	0	
Actual exceptional needs payments:						
Range of weekly averages	9	11-85	9	12-95	6	8-46
Mean		60		35		28
Median		60		38		33
Disregarded income	1	200	0		0	

(continued on p. 208)

TABLE A.1 (continued)

Group	1 Continuous soc. sec.		2 Mainly off work		3 Intermediate	
Number of families	9		10		14	
	No. of families	Amount weekly pence	No. of families	Amount weekly pence	No. of families	Amount weekly pence
DEDUCTIONS FROM SB SCALES:						
Voluntary savings:						
Range	6	50-150	1	150	0	
Mean		100				
Rent arrears						
Mean	2	19	3	43	0	
Wage stop:						
Range	0		4	47-306	3	*
Mean				141		+
Median				106		+
Notional exceptional needs payments:						
Range	9	20-45	10	20-35	14	15-45
Mean		26		27		24
Calculating errors detected	2	-100; +38	3	+	5	+
NET EFFECT ON FAMILIES' INCOMES OF ADDITIONS AND DEDUCTIONS:						
Better off than scales:						
Range	3	47-98	5	7-37	2	+
Mean		77		20		
Median		85		21		
Worse off than scales:						
Range	6	9-132	5	65-314	12	8-192
Mean		52		159		43
Median		42		172		25

* One 'voluntary unemployed' deduction of 75p.: one wage stop of £2.00; one uncertain wage stop.
+ Insufficient data prevented precise calculations.

Appendix B

PSYCHOLOGICAL INSTRUMENTS

1 PRINCIPAL COGNITIVE AND EDUCATIONAL TESTS

These are the items used to present normative and intergroup data in Figures 1-13. At each age there is a similar nucleus of such tests. The basis for selection was to cover the three broad areas of verbal ability (spoken and comprehended vocabulary), non-verbal ability ('g' loaded and perceptual-motor tasks), and reading (as the basic educational attainment). The items chosen were also to have, where possible, British norms or results from British representative samples, including samples of local (Midlands) children, to provide external reference points. All tests were administered individually.

The chosen tests will now be described.

Verbal: Here we chose vocabulary scales because these measures have high correlations with verbal intelligence tests such as the Stanford-Binet. At each age one vocabulary test used oral definitions by the child and the other used only a pointing response. This was in case socially handicapped children proved less able on the spoken than the comprehended vocabulary task, although in the event no such pattern emerged. The tests are:
6-7 years: Mill Hill Scale (oral) (Raven, 1958)
Examiner says single words, child gives meanings. There are separate norms for British boys in a large-scale study by Dunsden and Roberts (1957). We converted these to have a mean of 100, standard deviation 15.

10-11 years: Vocabulary subtest from Wechsler Intelligence Scale for Children (WISC) (Wechsler, 1949).
Administration similar to Mill Hill. Original norms are American. British comparison data, which we used to give the national estimates, are available from a survey of 100 boys and 100 girls aged $10\frac{1}{4}$ to $10\frac{3}{4}$, sponsored by the British Psychological Society (to be published by C.J. Phillips).

210 Appendix B

6-7 years: English Picture Vocabulary Test 1 (Brimer and Dunn, 1963).
Examiner shows multiple-choice set of four pictures; child points out the one corresponding to word spoken by examiner. Norms are British. Local data available from an unpublished study carried out in a Midlands borough by C.J. Phillips in 1969. The sample consisted of 2000 school children aged $6\frac{3}{4}$ to $7\frac{3}{4}$ years. We extracted the scaled (age corrected) scores of 1082 indigenous boys.

10-11 years: English Picture Vocabulary Test 2 (Brimer and Dunn, 1963).
Administration as for EPVT 1. Norms are British. Local data on fifty-two boys available from a random sample of Birmingham 11 year olds (Phillips and Bannon, 1968).

Non-verbal: We chose some 'g' loaded tests (sampling the general intelligence factor) and some others which have lesser 'g' loading and more perceptual motor element.
The 'g' loaded tests are:

6-7 years: Raven's Coloured Progressive Matrices (Raven, 1956).
Examiner shows a set of patterns making a design or matrix with a part missing. Subject selects the missing part from a multiple-choice set. Norms are British.

10-11 years: Raven's Coloured Progressive Matrices (Raven, 1956).
Administration as above. Norms are British. Local data available from same source as EPVT 2 above.

WISC Block Design Subtest (Wechsler, 1949).
One-inch cubes with coloured faces are used to construct patterns. Examiner shows model in blocks or on picture, child constructs. Norms and British data as for WISC Vocabulary subtest above.

The perceptual-motor tests are:

6-7 years: Draw-a-Man Test (Harris, 1963).
Subject draws a man, examiner evaluates on scoring standards. Norms are American.

10-11 years: Draw-a-Man Test (Harris, 1963).
Norms as above. Local data from same source as EPVT 2 above.

WISC Coding subtest (Wechsler, 1949).
Guided by a key, subject enters symbols under random sequence of digits. Norms and British data as for other WISC subtests.

Reading: At 6-7 years we chose a well-proven test which gave maximum discrimination among poor readers. It had been used in the National Child Development Study (Pringle et al., 1966), on which no normative data were published. At 10-11 years we chose two tests, one of which had local data, and the other to provide a measure of reading complete sentences, reflecting the broader reading skills required at this age.

211 Appendix B

6-7 years: Southgate Group Reading Test 1, Form C (Southgate, 1959).
From a multiple-choice list the child rings a word corresponding to that spoken by the examiner. Norms are British. Local data were available from same source as EPVT 1 above, except that since the test does not yield age-corrected scores the raw scores of forty-seven boys nearest in age to our sample were extracted.

10-11 years: Vernon Graded Word Reading Test (Vernon, 1938).
The child reads through graded list of uncorrected words. Norms are British. Local data from same source as EPVT 2 above, except that since the scores are not age-corrected an allowance was deducted from the local mean to compensate for the difference in mean age of that sample compared with ours.

NFER Reading Test AD (Watts, 1958), formerly Sentence Reading Test 1.
The child reads silently a sentence with a missing word and underlines an appropriate word from a multiple choice. Norms are British. No local data.

2 SUPPLEMENTARY TESTS, 6-7 YEAR OLDS

These were chosen to back up the major instruments and to assess some of the component skills of reading, writing and number.
   Psychologists will be familiar with the work of Jean Piaget on children's thinking. Piaget's tasks used materials such as bricks, modelling clay, and beads which might be familiar to the research boys in home or school. Although such items often show a heavy loading on the 'g' factor, it was felt that they might be less handicapping for the disadvantaged: this effect was not shown. At the time of our project there was a set of Piaget-type concept-development measures being researched by Professor E.A. Lunzer (subsequently the subject of a publication, 1970). We piloted some, and selected three as most likely to discriminate in our sample:
   Number: this item tests one-to-one correspondence (matching block to block to give 'the same number') and conservation of quantity, where the child shows whether he appreciates that a given number of blocks is invariant no matter what its spatial arrangement.
   Conservation of Linear and Circular Order: using coloured beads on laces, the child is asked to reproduce sequences, the most difficult of which have the examiner's lace reversed or in a figure of eight.
   Conservation of Weight: this uses plasticine and a set of scales. Various tasks determine whether the child knows that weight is conserved despite distortion of shape, or changed by addition even if external shape is unaltered.
In addition to the Lunzer-Piagetian tests there were the following:
   Ability to Copy Forms (ATCF): the copying of geometrical forms is known to be a component of the reading/writing skill and is also linked to general intelligence. The set of eighteen designs which are used were those of Graham et al. (1960), scored according

to criteria used in other research projects at the Centre for Child Study.

Motor Dots: this was included as a contrast to the complex skill of ATCF. The subject merely had to join up rows of dots (½ inch vertical separation between rows, ¼ inch between each pair of dots), by connecting each dot with one directly below it. There is a similar item in the Primary Mental Abilities test (Thurstone and Thurstone, 1953), having a high loading on a factor of hand-eye co-ordination. We added a second trial where the child was urged to go quickly. In the event, both trials, and the time and error scores within each, were highly correlated.

Sentence Repetition: developed for other projects in the Centre for Child Study, this item tests the subject's ability to repeat sentences of varying length and complexity immediately after hearing them.

British Intelligence Scale Induction Test: part of a test of the same name developed for the British Intelligence Scale (Ward and Fitzpatrick, 1970). The task required the child to complete series of coloured counters, and was thought likely to supplement the Matrices results.

3 OTHER TESTS USED ON 10-11 YEAR OLDS

New Junior Maudsley Inventory (Furneaux and Gibson, 1966) - see chapter 4.

Self-Concept Test - see chapter 4. The items were as follows, those marked 'E' being counted towards the Evaluation total, and those marked 'M' towards Motivation.

1 I occasionally get picked on by other boys. (E)
2 I have occasionally taken things which weren't mine. (E)
3 Boys get on with me. (E)
4 Sometimes I am bad tempered. (E)
5 I have a pleasant voice. (E)
6 I make plenty of mistakes in school work. (E)
7 I like sport and games. (M)
8 I have just as much money as other boys. (E)
9 I get bossed around by other boys. (E)
10 I am sometimes lazy. (E)
11 I am brave. (E)
12 I am pretty bad in at least one subject at school. (E)
13 I always help others when I can. (E)
14 I would rather go out to play than work in school. (E)
15 My clothes are a bit old. (E)
16 I want to come top, beat others. (M)
17 Some boys are stronger than me at my age. (E)
18 I get easily fed up with some people. (E)
19 I think hobbies like stamps, pets, and fishing are not much fun. (M)
20 My family gets on well with me. (E)
21 I sometimes annoy teachers. (E)
22 I sometimes feel less important than other boys. (E)
23 I usually watch the TV all the evening. (M)
24 I am good at games or sport. (E)

25 I want to leave school as soon as possible. (M)
26 I would rather save money than spend it. (M)
27 I am jealous. (E)
28 I am neat and clean. (E)
29 I always tell the truth. (E)
30 I am bored by reading. (M)
31 I am good at making models. (E)
32 I usually do better than others in school work. (E)

The Gibson Spiral Maze (Gibson, 1964): this is a psychomotor test where the subject draws a line through a spiral maze on a stiff card. The maze has circles at various points and obstacles to be avoided. The performance is timed and errors scored. It was originally included because Gibson reports that it discriminates between delinquents and non-delinquents, the former being quicker and less accurate. We do not report our results since we found that GWH, who tested two-thirds of the sample, produced better time and accuracy scores in his subjects than in those of the other examiners.

ADMINISTRATION AND SCORING

All tests were administered individually in the boys' own schools, except in the case of three 6-7 year old focus boys and one 10-11 year old, who attended school so infrequently that it was necessary to bring them from their homes to the Centre for Child Study. This seemed to have an adverse effect. The test batteries took about an hour and a half.

The order of administration was as follows:

6-7		*10-11*	
1	ATCF	1	WISC Block Designs
2	EPVT 1	2	WISC Coding
3	Draw-a-Man	3	WISC Vocabulary
4	Mill Hill Vocabulary	4	Vernon Reading
5	Coloured Progressive Matrices	5	NJMI
6	Southgate Reading Test	6	Draw-a-Man
7	Motor Dots	7	Coloured Progressive Matrices
8	BIS Induction	8	Gibson Spiral Maze
9	Lunzer Concept Tests:	9	EPVT 2
	(i) Number	10	NFER Reading Test
	(ii) Weight	11	Self-Concept Test
	(iii) Order		
10	Sentence Repetition		

The general principle was to have items requiring no verbal response at the beginning, to act as a comfortable introduction, and to vary the mixture thereafter.

All items with an element of subjectivity in the scoring were rescored by GWH if another examiner had administered them. Checks were performed to ascertain whether his scoring of the most subjective tests, ATCF and Draw-a-Man, was reliable (this was confirmed, all inter-scorer coefficients being .90 or higher), and whether the means from his scoring were significantly different from those of a second person. The 10-11 year old Draw-a-Man test showed significantly higher mean scores for GWH than from the other scorer. Probably this scale as reported at 10-11 years has means about one or two scaled score points high, but it is known to be subject to inter-examiner differences.

BEHAVIOUR-RATING SCALES

There were separate scales for the two age-groups. The one used for the older boys was a longer and more developed version of the other, so there is much similarity of content. The earlier version had items grouped under four headings: Anxiety, Behaviour Problems, Maturity (later Competence), and Introversion-Extraversion. The later version used most of these items or slightly altered versions, together with additional ones. They were grouped not under subscale headings but under the headings 'The Child and Adults', 'The Child and Other Children', 'The Child in School' and 'General Trends'. The detailed rationale of the scales is given elsewhere (Herbert, 1972, 1973, 1974a) and the items are given in full by Herbert (1972) and Phillips, Wilson and Herbert (1972). Each item is in the form of a five-point continuum with the two extremes and a midpoint defined by wording. The wording was kept neither too abstract nor too narrowly behavioural, the items being selected as sampling characteristics readily observable in school. The following is an example:

Desire for sympathy from staff

X	X	X	X	X
Always comes up with minor hurts or tales of being wronged by others		Does not make an unnecessary fuss		Keeps his grievances, if any, away from adults' sight

Each item had a title like this one, to orientate the rater to its subject matter. The teacher was asked to ring one of the Xs. A score from one to five could be given to the points on the continuum.

The allocation of items to subscales in the developed version used for the 10-11 year olds was decided by factor analysis of the scores from 141 boys covering a wide social range, half of whom were not in the present research project. Items were scored in only one subscale - that of the factor in which they had their highest loading or, in the few cases of ambiguity, the scale in which they scored to fit best. All the items in a scale loaded 0.35 or over on the relevant factor.

The scoring of subscales A, C, D and E at 10-11 years was simply by adding the relevant item scores (1 to 5) into a total. The B(I) and B(E) subscales were made up from a single group of items, those with a common strand of introverted versus extraverted behaviour difficulties. Instead of being scored 1-5 toward a single 'behaviour problems' total, the poles of these items were used to make each a separate subscale. Starting with the midpoint of the original item, the B(E) subscale used the points out to the extraverted extreme, and the B(I) subscale the points out to the introverted extreme. The points were assigned weightings derived from a study where independent judges were asked to assess the severity of the problems depicted by the wording in each item.

The early version of the scale, used on the 6-7 year olds, was scored as follows. The A and C dimensions were confirmed by independent factor analysis (Hague, 1970), so the groups were used as given by the headings in the original form. The B(I) and B(E) scales were made up from those items, some Behaviour Problems, some Introversion-Extraversion, which were the same as or very like some in the scale used on the older boys. The weightings for the points on the items were decided in the same way as for the 10-11 year olds. This group of items was thus parallel to the group used on the older boys. In Hague's factor analysis it had loaded two factors but they were not distinct, being correlated .64.

Limitations of space prevent us from giving the items in full. Abbreviated item descriptions are given below. Although we consistently put the 'high' pole of the item on the left, the orientation was randomized in the actual rating forms.

A Subscale

6-7 years and 10-11 years

Tension Worried expression	versus	Never looks bothered
Mannerisms Shows frequently	versus	Free of such habits
Moods Has times when depressed, withdrawn	versus	Very sunny character
Reaction to change in school Upset	versus	Not very moved
Difficulty settling in school Uneasy or complains	versus	Very willing, unaware of separation from home

6-7 years only

Contacts with peers Uninhibited	versus	Inhibited (E subscale for 10-11 years)
Dependency Frequently clings	versus	Never shows (D subscale for 10-11 years)
Trustfulness Suspicious, wary	versus	Trustful in naive way (D subscale for 10-11 years)
Fearfulness Many fears and apprehensive	versus	Little or no sense of danger (B subscales for 10-11 years)

10-11 years only

Attitude when asked question Quickly bothered	versus	Very little concern

216 Appendix B

B(I) and B(E) Subscales

6-7 and 10-11 years

Noisiness Noisy, crashes, shouts	versus	Excessively quiet
Resentful Resents adult correction, intervention	versus	Always malleable
Dominance Dominates peers	versus	Recessive, submissive
Temper Explosive after minor provocation	versus	Rarely seen
Sharing Grabs, or always in conflicts	versus	Allows others to take with little protest
Playing up 'Silly', provocative	versus	Subdued
Defiance Answers back, resistive	versus	Under-assertive
Physical Aggression Highly aggressive	versus	Under-aggressive
Activity Level Overactive	versus	Slow, deliberate

6-7 only

Attention-seeking Obvious and active	versus	Retiring (D subscale for 10-11 years)
Impulsive Lack of forethought	versus	Very cautious
With visitors Pushes self forward	versus	Withdraws (D subscale for 10-11)

10-11 only

Showing off Shows off before group	versus	Will not do this
Restraint Starts asking before instructions finished	versus	Has to be encouraged
Self-Confidence Very sure, perhaps oversure	versus	Very cautious
Fidgets Doesn't sit still for long	versus	Sits very still
Excitability Easily very excited	versus	Never
Stealing Frequent	versus	Never
Lies Frequent	versus	Never

C Subscale

6-7 and 10-11 years

Independence Copes well on own	versus	Often needs help
Classroom jobs Spontaneously does well	versus	Cannot be relied on
Carrying out instructions Anticipated, grasps well	versus	Poor grasp

Perseverance High, in spite of difficulties   *versus*   Low, even in minor difficulties
Choices Mature and deliberate   *versus*   Indecisive, or limited
When fails Self-composed   *versus*   Upset or underconfident
Helping others Useful, resourceful   *versus*   No help or too much

6-7 only

Dressing Manages all items   *versus*   No initiative or careless
Social contacts Many, or mature choice of few   *versus*   Unpopular or 'lame duck'

10-11 only

Motivation Great efforts   *versus*   Marked lack of effort
Imagination High   *versus*   Low
Tidiness Neat, orderly   *versus*   Untidy or destructive
Jobs outside classroom Can be entrusted   *versus*   No certainty of success

D Subscale (10-11 only)

Demand for praise Persistent   *versus*   Seems indifferent
Communication Tells about minor detail   *versus*   Prefers not to
Desire for sympathy Tells of minor grievances   *versus*   Keeps away from adult sight
Dependency Frequently clings   *versus*   Never shows (A subscale for 6-7)
Attention seeking Obvious and active   *versus*   Retiring (B(I) and B(E) for 6-7)
With visitors Pushes self forward   *versus*   Withdraws (B(I) and B(E) for 6-7)

E Subscale (10-11 only)

Group work Welcomes   *versus*   Prefers individual work
Group play Invites others   *versus*   Rejects or ignores
Friendships Desires a lot   *versus*   Doesn't seem concerned
Trustfulness Suspicious, wary   *versus*   Trustful in naive way (A subscale for 6-7)
Jealousy Low   *versus*   High
Social Contacts Relaxed   *versus*   Inhibited

Conduct Scale (West, 1969)

This scale had nine items summarized by Gibson (1964) as follows; together with scoring weights derived from a principal components analysis.

Appendix B

		Weight
1	Truanting	5
2	Difficult to discipline	5
3	Not concerned to be a credit to his parents	5
4	Noticeably dirty and untidy	5
5	Lazy	4
6	Difficulty with peers due to aggressiveness, etc.	2
7	Seriously distractable	2
8	Not specially good at anything	1
9	Outstandingly bad in at least one subject	1
	Max. score	30

Dr Gibson kindly communicated the scoring system to us.

Appendix C

TABLES OF PSYCHOLOGICAL DATA

TABLE C.1 Raw score means and standard deviations for cognitive tests: 6-7 year olds

	Controls Low SH		Controls Moderate SH		Focus	
	Mean	SD	Mean	SD	Mean	SD
English Picture Vocabulary Test 1	18.4	5.3	11.9	6.1	8.8	5.5
Mill Hill Vocabulary Test	11.5	4.2	8.6	2.7	6.4	3.2
Raven's Coloured Progressive Matrices	15.8	3.2	14.5	2.7	12.2	3.8
Southgate Reading Test 1, Form C	9.5	5.7	9.7	4.7	5.0	4.5
Harris Draw-a-Man Test	18.1	5.8	18.6	4.4	14.4	6.9
Ability to Copy Forms	13.5	2.0	12.5	1.9	10.3	3.0
Lunzer: Number concepts	3.4	2.7	2.5	2.5	1.5	1.8
Lunzer: Linear and Circular Order	7.6	4.0	5.5	3.7	2.8	3.3
Lunzer: Conservation of Weight	2.9	1.5	3.1	1.7	1.5	1.6
Dots: first trial, time (secs)	38.6	9.2	42.2	8.7	48.4	13.9
Dots: first trial, errors	3.8	3.1	6.1	7.9	7.0	7.1
Dots: second trial, time (secs)	25.9	5.8	29.0	6.8	35.8	16.7
Dots: second trial, errors	8.0	4.1	9.9	7.0	13.4	8.6
Sentence Repetition	10.0	1.9	9.1	1.8	8.2	2.6
British Intelligence Scale Induction	4.5	1.6	3.9	1.7	2.9	1.8
N =	30		26		28	

TABLE C.2 Raw score means and standard deviations for cognitive tests: 10-11 year olds

	Control Low		Control Moderate SH		Focus	
	Mean	SD	Mean	SD	Mean	SD
WISC Vocabulary Subtest	35.9	6.1	32.4	6.9	28.8	6.6
English Picture Vocabulary Test 2	22.0	7.0	19.2	7.9	15.2	7.0
WISC Coding Subtest	35.7	8.3	33.5	6.1	28.7	7.4
WISC Block Design Subtest	23.2	12.1	17.1	9.7	14.0	8.6
Raven's Coloured Progressive Matrices	26.6	5.2	24.3	4.7	22.5	5.1
Harris Draw-a-Man Test	37.5	8.7	35.6	8.3	30.9	10.0
Vernon Graded Word Reading Test	40.1	28.5	30.9	22.2	27.2	25.4
NFER Sentence Reading Test AD	17.8	11.4	15.3	9.7	12.9	11.6
N =	31		30		29	

TABLE C.3 Raw and standardized score means and standard deviations - behaviour-rating scales

		Control Low		Control Moderate SH		Focus	
		Mean	SD	Mean	SD	Mean	SD
A: 6-7 years	Raw	21.1	4.8	24.1	6.2	23.5	6.8
	Standard	48.3	8.8	53.2	11.3	53.6	12.3
A: 10-11 years	Raw	13.8	4.7	15.9	4.1	16.7	3.8
	Standard	48.3	10.5	52.5	8.6	54.6	8.9
B(I): 6-7 years	Raw	8.5	7.9	10.7	8.1	14.1	10.3
	Standard	49.2	10.3	51.5	9.5	56.5	13.2
B(I): 10-11 years	Raw	17.4	12.6	16.4	11.4	17.2	15.0
	Standard	50.5	10.4	49.3	9.3	50.3	12.2
B(E): 6-7 years	Raw	5.6	8.2	6.3	9.1	8.7	11.6
	Standard	49.7	9.7	50.5	10.9	53.3	13.6
B(E): 10-11 years	Raw	6.2	7.8	6.2	8.2	10.3	15.8
	Standard	50.1	9.7	49.9	10.5	54.5	18.9
C: 6-7 years	Raw	29.9	6.3	27.7	7.5	20.8	7.0
	Standard	51.0	9.3	48.0	11.0	37.4	10.5
C: 10-11 years	Raw	39.2	9.4	34.7	9.1	31.8	10.2
	Standard	52.0	10.0	47.3	9.5	43.8	10.8
D: 10-11 years	Raw	15.8	3.8	16.6	3.7	18.5	4.7
	Standard	49.0	10.0	51.4	9.9	56.0	12.7
E: 10-11 years	Raw	24.5	3.3	24.0	4.0	22.4	4.1
	Standard	50.4	8.8	49.5	11.6	44.6	11.9
Conduct: 10-11 years	Raw	6.5	4.2	9.9	5.6	13.8	6.7
	Standard	47.2	8.2	53.8	11.2	61.7	13.4

The standard score data are given because the F ratios quoted in chapter 4 used these statistics. Their mean of 50 and standard deviation of 10 refer to the population from which the control boys were drawn.

TABLE C.4 Combined correlation matrix for control groups: 6-7 year olds

	1	2	3	4	5	6	7	8	9	10	11	12	13	14	15
1 ATCF	–														
2 EPVT 1	40	–													
3 Draw-a-Man	32	13	–												
4 Mill Hill	25	39	36	–											
5 Matrices	34	23	41	16	–										
6 Dots first speed	-11	-22	-17	-06	-22	–									
7 Southgate	32	24	15	48	09	-37	–								
8 Behav. rating:A	-15	-09	06	16	-03	20	03	–							
9 Behav. rating: (BI)	-26	-29	-06	06	00	23	-01	28	–						
10 Behav. rating: B(E)	10	14	05	-12	-12	-19	-11	21	-54	–					
11 Behav. rating:C	20	25	-05	21	08	-22	37	-52	-08	-30	–				
12 Lunzer: Number	-01	13	29	32	28	-11	18	-20	11	-16	31	–			
13 Lunzer: Order	18	11	44	41	21	-29	27	-22	06	-24	22	34	–		
14 Lunzer: Weight	11	31	15	35	27	-09	20	-17	02	-08	23	34	26	–	
15 Sentence Repetition	01	-01	00	15	-10	-14	48	12	12	-03	22	04	12	22	–
16 BIS Induction	24	21	17	25	00	-08	29	-14	-20	-12	31	07	18	11	19

Decimal points omitted.
Critical values of r
p = 0.05   0.01   0.001
r   26   34   43
Notes
1 Variables were converted to our own internally standardized scores before correlation. The procedure incorporated an age correction which partials out the effect of age.
2 Due to the high intercorrelation of all forms of the Dots test, the first trial time score only was used.

222 Appendix C

TABLE C.5 Combined correlation matrix for control groups: 10-11 year olds

	1	2	3	4	5	6	7	8	9	10	11	12	13	14	15	16	17	18	19
1 EPVT 2	—																		
2 WISC Vocabulary	74	—																	
3 WISC Block Design	31	34	—																
4 WISC Coding	13	19	31	—															
5 Matrices	47	35	66	31	—														
6 Draw-a-Man	34	32	19	-06	28	—													
7 Vernon Reading	56	64	41	40	53	28	—												
8 NFER Reading	52	58	44	42	53	26	95	—											
9 NJMI Extraversion	-10	11	-23	-12	-12	00	04	-01	—										
10 NJMI Neuroticism	-01	08	04	-07	-09	04	-02	-13	-28	—									
11 NJMI Lie	06	-16	-22	-13	-11	-06	-10	-01	04	-53	—								
12 Self-Concept Evaluation	13	-10	07	15	25	09	11	16	10	-45	62	—							
13 Self-Concept Motivation	24	36	04	22	10	06	31	30	18	-18	15	25	—						
14 Behaviour Rating A	-28	-28	-17	-35	-39	-04	-37	-35	-28	07	-02	-16	-18	—					
15 Behaviour Rating C	54	55	34	41	41	08	54	57	-03	-24	08	17	35	-40	—				
16 Behaviour Rating D	09	-07	-14	-17	-23	00	-16	-21	39	-11	-02	00	06	00	-03	—			
17 Behaviour Rating	04	-09	-06	16	-13	-09	04	05	26	02	-17	-18	05	-32	10	48	—		
18 Behaviour Rating B(I)	03	-04	-07	-11	03	-02	-06	02	-35	-04	19	23	-13	26	12	-42	-54	—	
19 Behaviour Rating B(E)	-01	12	-14	01	01	-02	10	-04	33	16	-23	-24	17	-14	-27	10	-58	—	
20 Conduct	-39	-33	-27	-41	-25	-05	37	42	10	18	-08	-14	-16	32	-79	07	-16	-23	56

Decimal points omitted
Critical values of r
p = 0.05 0.01 0.001
r = 25   33   41

Appendix D

INTERVIEW SCHEDULES: THE FAMILY

INTERVIEW 1

Introduction. I am glad you have agreed to help us. It's not always easy to make time when you have such a lot to do. I have children of my own and although they give a lot of pleasure, it can be quite a job at times looking after them. I'm interested to know how you manage with yours and also what you do nowadays when prices go up and wages don't. Has X said anything about the Nursery group? What do you think about it? Does it make much difference to you on the morning X is there?

1 How many children have you?

2 How old are they?

Name	M	F	Age	School/Job

3 Are they all still living with you?
   Are any of them married?

4 It can be quite a rush in the mornings. Do you always find it possible to get them up for a breakfast before they have to be out? *Verbatim*

5 What about getting them ready for school? Do you have to do most of this or can they do quite a bit for themselves? *Verbatim*

6 Do they have far to go to get there? (NB Different schools may be attended.) *Verbatim*

7 Are there children who have left school? Where do they work?
   For each child: (a) How long have they worked there?
                   (b) Have they had any other job?
                   (c) Do you mind telling me why they changed?
                   (d) Are jobs difficult to find?

Appendix D

Name	Job	Since when	Previous job(s)	Comments

8 What does your husband do?
   (b) How old is he?
   (If has job): prompt for F.'s hours of work, regularity of employment, reasons for changing jobs.
   (If has no job) prompt as to last job, reasons for unemployment, previous work pattern, what he does during the day and attitude to unemployment

F. at work :	hours
Normal day	
Shifts	
Mainly nights	
Casual (sporadic hours)	
Other	

9 We would be glad if you could tell us how you manage? Obviously you have an awful lot to get through. For instance, what did you do yesterday?

Mother's Comments

	Mother	Pre-school children	Other children	Father
Breakfast				
Lunch/Dinner				
Tea/Dinner				

10 Is this your general routine or does it vary a lot from day to day? *(If it does, prompt for in what ways.)* *Verbatim*

11 Are there things you feel you must get done every day? *Verbatim*

12 Do the children or your husband do anything around the house? *(Prompt for specific tasks.)* *Verbatim*

13 Do they do this regularly or just occasionally?

14 You mentioned shopping, who usually does it? *Verbatim*

15 With so many to feed, do you have to shop often?

16 Do you make a shopping list or do you find it easier to decide what you want when you get there? *Verbatim*

17 Are the shops you like handy or is it a bit of an expedition getting there? *Verbatim*

18 Do you enjoy shopping? *Verbatim*

19 When it comes to lunch time, who is usually home? *Verbatim*

Appendix D

20 Are the children fussy about what they have, or is it easy to provide variety? *Verbatim*
21 What did they have today?
22 Do any of them stay for school dinners? *(Explore reasons for staying or not staying.)*
23 Do you have any difficulty in getting the children who do come home, back to school in time? *Verbatim*
24 What do the children who are not at school do? *(Prompt for mornings and afternoons.) Verbatim*
25 Do you ever manage to go out or have friends in?
    Is it easier to manage this in the afternoons or the evenings?
26 By teatime, is everyone usually in, or do you find the younger ones are hungry before your husband and some of the older ones are in, so that you have to give them their tea first?
    Is there any time when you are all in together and can sit down as a family for a meal?
27 By teatime, lots of mothers feel they've had a full day, do you find you can have a rest and maybe a talk with your family or does it not work out that way?
    (Interviewer note)
    (a) if sufficient chairs for all of family to sit down together
    (b) if present at mealtime, who serves out food and organizes meal?
28 Supposing one of the children wanted something to eat in between meals, would you let him? Can he help himself or must he ask you first?
29 After tea, in the evenings, who is usually in?
30 What sort of things do you do at home?
31 In some families, TV causes a lot of arguments. Sometimes people want different programmes, do you find this? *(Find out who actually decides what shall be watched and if once on, TV stays on.)*
32 Getting children to do homework in the evenings can be a problem too. Do any of your children have homework to do? Is it easy to get them to do it?
33 Does anyone have any regular nights out?
34 What sort of things do they like to do?

Family Member	Activities

35 Now what about bed-time? Can you tell me what time they went to bed last night? *(Prompt - Is this the time they usually go and specifically for pre-school child and child under focus.)*

36 Do you usually send them all off together? Or do they each have a special time?

37 Does anyone help you to get them off?

38 Do all of them have to share a room?
    Does this make it harder for them to settle down at night?

39 You've told me a lot about what happens on weekdays that has given a very helpful picture of your life. Are the weekends much different? *(Prompt for differences, e.g. getting up times, less housework, children going to bed later.)*

40 Sometimes having the children around all day can be a bit of a strain. Do you find this at weekends?

41 Do you feel the same way during school holidays?

42 Is it possible at weekends and holidays to get out all together?

INTERVIEW 2

Introduction. What you told me last time was very helpful. We'd like to know a bit more about how you manage, for instance when there is trouble whom can you turn to for help? Sometimes it's not easy to get someone to help because there is no one close by you know well enough.

Area

1 Have you always lived in this area?
   YES/NO

2 *If YES,*
   Were both you and your husband brought up here?

   If NO,
   Where did you and your husband grow up?
   When did you leave there?
   Have you moved about quite a bit or did you come straight here?
   Leaving old friends can be painful, do/did you find it lonely when you left? Have/had you a special friend there?

Relatives

3 What about your relations, do they find it easy to come to see you? *(Prompt for both M.'s and F.'s.)*

4 So you do/don't see a lot of them?

5 Are any of them able to come to help you if you were ill or had a baby or something else caused an upset?

6 Many people find they look to other people in their family for help and advice, is there anyone in particular amongst your relatives to whom you would go?

7   Is this the person you get on best with and feel closest to in the family?
8   Do any of your relatives rely on you for help?

Neighbours

9   What about your neighbours? If you found it difficult to get out because the children were ill or if you found you had run out of tea or something, would you be able to go to them?
10 Do you find they ask you to do the same sort of things?
11 When you moved in here, did you find it easy to settle in?
12 Do you think it's the people around here or other things that make you feel that way?

Friends

13 Do most of your friends live near here or are they in Y? *(If have moved.)*
14 In some families, men have their friends and the women theirs, is this the way it is in your family or do you have most of your friends in common?
15 Are you able to see them often?
16 It's good to get a night out. If you wanted to go out with your friends, would you be able to get someone always to look after the children or can you trust one of the older ones? *(Prompt for who would come.)*
    Do you do the same in return for them or do you do something else for them?
    Do you always tell N. when you're leaving him, or do you find it easiest to slip off without him knowing?

Crises

17 When someone is ill in the family, it's not always easy to decide when to call in the doctor or when to go to see him. How do you do this? Do you find it easy to get the doctor to come?
18 Does anyone in the family suffer much from illness? *(Prompt so that all the family are covered.)*
19 It would be helpful if you could tell me what sort of illness you have had to cope with. *(Prompt for whether M. or children have had to go into hospital.)*
20 Some children always seem to be falling sick, are any of yours like this?
21 So on the whole illness in your family:
    (a) happens a lot?
    (b) doesn't happen very often?

228   Appendix D

22 Just as some children seem always to be getting colds, others have accidents. Have you found this with any of yours? *(Prompt for details.)*

23 With the younger children, have you had much to do with the Health Visitor?

24 Are the times the clinic is open easy for you to get along? *(Prompt for whether the clinic is visited or not.)*

25 Now, what about you, if a father or mother have a serious illness or accident, it can cause quite an upheaval, have you ever had to manage in these circumstances? *(Prompt for details.)*

26 *(If not previously covered.)* Did you find that when you were having your children, you could make arrangements for the older ones easily? How did you do this?

27 By and large you seem to have been lucky/unlucky as far as health goes. Has anything ever happened so that you've really felt at the end of your tether?

28 Do you mind telling me about it?

29 Most of us when this sort of thing happens feel we need to go to someone else for help and support. Who would you go to?

30 *(If applicable.)* You've mentioned friends and relatives helping you. Would you ever go to a social worker or someone like that? Can you give me an example of something you have gone to them for?

(Interviewer to note whether a neighbour is ever present during any interview.)

INTERVIEW 6   (FINAL)

1 *Housing*

House	Flat	Rooms	Self-contained	Shared
Furnished	LA	Own	Rented from landlord	
Terraced	Semi-detached	Backhouse		

Rooms in use: BR   LR   Kitchen   Scullery   Bath
                Toilet   in/out   shared

Hot water: No. of taps   Hand basins

Garden: Yard   Play space

Lived here since:

House due for demolition?

Are you on housing list?

2 *Possessions*

TV   Radio   Record player   Phone   Fridge
Washing Machine   Vacuum cleaner   Car   Bike   Scooter
Pets   Other:

3 *Family income and occupations*
    Father
    Mother
    Sibs

    Insurance benefit
    SB
    FA
    Other:

4 *Outgoings*
    Rent   Arrears
    HP   'Club'

        Av. weekly:
    Electricity   (Debt?   )
    Gas   (Debt?   )
    Coal   (Debt?   )
    Other fixed expenditure:
    Court order?

5 *Means-tested benefits*
    Rent rebate: Applied?   Granted?
                    If not, why?

    Rate rebate: Applied?   Granted?
                    If not, why?

    Milk for under-5s   Free? Reduced?
    How much milk do you take daily?

    Prescription charges: Applied for repayment? Exemption?
    Glasses
    Teeth
                    If not, why?

    Educational grants for clothing:
                    Applied?   Granted?
                    If not, why?

    Educational maintenance grant:

6 *Family planning*
    Interested?
    Agency?
    Would you prefer domiciliary FP?

7 *Family history*
    Mother   School:
                Further education:
                Training:
                Work history:
                Job satisfaction:

    Father   School:
                Further education:
                Training:
                Work history:
                Job satisfaction:

7 *Family history (continued)*
    Previous marriage: (Mother?   Father?)
    Date: Children: Staying with:

8 *Parents joint life*
    Age when married: Mother: Father:
    Date of marriage:

   Phase I (before arrival of first child)
      Work?
      Income?
      Housing?

   Phase II (children)
      Mother's age:
      Work?
      Income?
      Housing?

9 *Aspirations for children*
    Mother   Father

Appendix E

INTERVIEW SCHEDULES: THE CHILDREN

The following questions from the Nottingham interview schedules have been used:

1 *Interview 3* (from 'Four Years Old in an Urban Community'):
Independence: 6, 9, 10, 11. Aggression in play: 17-22. Autonomy in play: 29, 30, 34. Personal habits: 46-9. Bedtime: 50-54, 65-9, 70. Toilet training: 71-9. General discipline: 80-103, 106. Father participation: 107-10.

2 *Interview 4* (from 'Seven Years Old'):
Personality: 5-15, 51, 55, 60-71, 77-81. Discipline and do's and don'ts: 153-60, 162-4, 83, 85, 43-5. School and activities: 95, 98-100, 103-7, 109, 112-17, 120, 121, 125, 128, 129, 134, 139, 143, 147-50, 46, 17, 18, 31-5, 21-4, 87-94.

3 *Interview 5* (from 'Eleven Years Old'):
Personality: 1-8, 10, 13, 14. Social relationships: 15-17, 21-4, 31, 32. Stress: 50, 55, 57-60, 62-5. Activities: 71, 79, 81-7, 90-1. School: 95, 96, 106, 107, 110-13, 115. Family interaction: 122, 126, 128, 130-2, 134, 137, 139, 140, 142, 143, 146. Conflict: 149-51, 153, 154, 156-61, 164, 166-74, 176-84, 186-91.

REFERENCES

ABEL-SMITH, B. and TOWNSEND, P. (1965), 'The Poor and the Poorest', London, Bell.
ASKHAM, J. (1969). Delineation of the lowest social class, 'J. Biosoc. Science', 1, 327-35.
ATKINSON, A.B. (1973), Low pay and the cycle of poverty, in Frank Field (ed.), 'Low Pay', London, Arrow Books.
ATKINSON, A.B. (1974), Poverty and income inequality in Britain, in Dorothy Wedderburn (ed.), 'Poverty, Inequality and Class Structure', Cambridge University Press.
BARNES, J. and LUCAS, H. (1975), Positive discrimination in education: individuals, groups, and institutions, in J. Barnes (ed.), Educational Priority, Vol. 3, 'Curriculum Innovation in London's EPAs', London, HMSO.
BECKER, H. (1963), 'Outsiders: Studies in the Sociology of Deviance', New York, Free Press.
BERGER, J. and LUCKMANN, T. (1967), 'The Social Construction of Reality', Harmondsworth, Penguin.
BERNSTEIN, B. and DAVIES, B. (1969), Some sociological comments on Plowden, in R.S. Peters (ed.), 'Perspectives on Plowden', London, Routledge & Kegan Paul.
BIRCH, H.G., RICHARDSON, S.A., BAIRD, D., HOROBIN, G. and ILLSLEY, R. (1970), 'Mental Subnormality in the Community', Baltimore, Williams & Wilkins.
BLURTON-JONES, N. (ed.) (1972), 'Ethological Studies of Child Behaviour', Cambridge University Press.
BODMER, W.F. (1972), Race and IQ: the genetic background, in K. Richardson and D. Spears (eds), 'Race, Culture, and Intelligence', Harmondsworth, Penguin.
BOXALL, M. (1973), Multiple deprivation: an experiment in nurture, 'Occasional Papers of the Division of Educational and Child Psychology of the British Psychological Society', No. 2, Spring 1973.
BRENNAN, M. (1972), Medical characteristics of children supervised by the local authority social services department, 'Policy and Politics', 1, 255-66.
BRIMER, M.A. and DUNN, L.M. (1963), 'Manual for the English Picture Vocabulary Tests', Slough, National Foundation for Educational Research

BRUNER, J.S. (1966), 'Towards a Theory of Instruction', Cambridge, Mass., Harvard University Press (page references to paperback edition 1968, New York, Norton).
BRUNER, J.S. (1971), 'The Relevance of Education', Cambridge, Mass., Harvard University Press (page references to paperback edition 1973, New York, Norton).
BULLOCK, Sir Alan (1975), 'A Language for Life', London, HMSO.
BURT, C. (1958), The inheritance of mental ability, 'Amer. Psychol.', 13, 1-15.
BURT, C. (1966), The genetic determination of differences in intelligence: A study of monozygotic twins reared together and apart, 'Brit. J. Psychol.', 57, 137-53.
BUTLER, N.R. and BONHAM, D.G. (1963), 'Perinatal Mortality', London, Churchill Livingstone.
CANE, B. and SMITHERS, J. (1971), 'The Roots of Reading', Slough, National Foundation for Educational Research.
CATTELL, R.B. (1963), Theory of fluid and crystallised intelligence: A critical experiment, 'J. Educ. Psychol.', 54, 1-22.
CENTRAL ADVISORY COUNCIL FOR EDUCATION (1963), 'Half Our Future' (Newsom Report), London, HMSO.
CENTRAL ADVISORY COUNCIL FOR EDUCATION (1967), 'Children and their Primary Schools' (Plowden Report), London, HMSO.
CHAZAN, M. and JACKSON, S. (1971), Behaviour problems in the infant school, 'Journal of Child Psychology and Psychiatry', 12, 191-210.
CHIEF MEDICAL OFFICER OF THE DEPARTMENT OF EDUCATION AND SCIENCE (1972), 'The Health of the School Child 1969-70', London, HMSO.
COMMITTEE ON ABUSE OF SOCIAL SECURITY BENEFITS (1973), 'Report' (Fisher Report), London, HMSO.
CULYER, A.J., LAVERS, R.J. and WILLIAMS, A. (1972), Health indicators, in A. Shonfield and S. Shaw (eds), 'Social Indicators and Social Policy', London, Heineman.
DANIEL, W.W. (1974), 'National Survey of the Unemployed', London, Political & Economic Planning.
DAVIE, R., BUTLER, N. and GOLDSTEIN, H. (1972), 'From Birth to Seven', London, Longman.
DEPARTMENT OF EMPLOYMENT (1971), New Earnings Survey 1970, Part 3, Analysis by Region, 'D.E. Gazette', LXXIX, 1, London, HMSO.
DEPARTMENT OF EMPLOYMENT (1971), 'Family Expenditure Survey 1970', London, HMSO.
DEPARTMENT OF HEALTH AND SOCIAL SECURITY (1972), 'Annual Report', London, HMSO.
DEPARTMENT OF HEALTH AND SOCIAL SECURITY, SUPPLEMENTARY BENEFITS COMMISSION (1971), 'Supplementary Benefits Handbook', London, HMSO.
DOUGLAS, J.W.B. (1964), 'The Home and the School', London, MacGibbon & Kee.
DOUGLAS, J.W.B., ROSS, J.M. and SIMPSON, H.R. (1968), 'All Our Future', London, Peter Davies.
DUNN, L.M. (1959), 'Manual for the Peabody Picture Vocabulary Test', Minneapolis, American Guidance Service.
DUNSDON, M.I. and ROBERTS, J.A.F. (1957), A study of the performance of 2000 children on four vocabulary tests. II. Norms, with some observations on the relative variability of boys and girls, 'Brit. J. Stat. Psychol.', 10, 1-16.
EDWARDS, J., LEIGH, E. and MARSHALL, T. (1970), 'Social Patterns

in Birmingham', Centre for Urban and Regional Studies, University of Birmingham.
EVERSLEY, D. (1973), Problems of social planning in inner London, in D. Donnison and D. Eversley (eds), 'London: Urban Patterns, Problems, and Policies', London, Heineman.
EVERSLEY, D. (1975), Landlords' slow goodbye, 'New Society', 16 January 1975.
EYSENCK, H.J. (1971), 'Race, Intelligence, and Education', London, Temple Smith.
EYSENCK, S.B.J. and EYSENCK, H.J. (1963), On the dual nature of extraversion, 'Brit. J. Soc. Clin. Psychol.', 2, 46-55.
FERGUSON, N., DAVIES, P., EVANS, R. and WILLIAMS, P. (1971), The Plowden Report recommendations for identifying children in need of extra help, 'Ed. Res.', 13, 210-11.
FERGUSON, N. and WILLIAMS, P. (1969), The identification of children needing compensatory education, in Schools Council Project in Compensatory Education, 'Children at Risk', Occasional Publication No. 2.
FIELD, F. (1973), 'Low Pay', London, Arrow Books.
FIELD, F. (1975), 'Unemployment: the Facts', London, Child Poverty Action Group.
FIELD, F. and TOWNSEND, P. (1975), 'A Social Contract for Families', London, Child Poverty Action Group.
FISHER COMMITTEE, FISHER REPORT, see Committee on Abuse of Social Security Benefits.
FURNEAUX, W.D. and GIBSON, H.B. (1966), 'The New Junior Maudsley Inventory Manual', University of London Press.
GANS, H.J. (1970), Poverty and culture: Some basic questions about methods of studying life-styles of the poor, in P. Townsend (ed.), 'The Concept of Poverty', London, Heineman.
GIBSON, H.B. (1964), The spiral maze - a psychomotor test with implications for the study of delinquency, 'Brit. J. Psychol.', 55, 219-25.
GIBSON, H.B. (1967), Self-reported delinquency among schoolboys and their attitudes to the police, 'Brit. J. Soc. Clin. Psychol.', 6, 168-73.
GIBSON, H.B. (1969), The significance of 'Lie' responses in the prediction of early delinquency, 'Brit. J. Educ. Psychol.', 39, 284-90.
GOLDTHORPE, J.H. (1974), Social inequality and social integration in modern Britain, in Dorothy Wedderburn (ed.), 'Poverty, Inequality and Class Structure', Cambridge University Press.
GOLDTHORPE, J.H., LOCKWOOD, D., BECHHOFER, F. and PLATT, J. (1969), 'The Affluent Worker in the Class Structure', Cambridge University Press.
GRAHAM, F.K., BERMAN, P.W. and ERNHART, C.B. (1960), Development in preschool children of the ability to copy forms, 'Child Develop.', 31, 338-59.
GREVE, J., PAGE, D. and GREVE, S. (1971), 'Homelessness in London', Edinburgh, Scottish Academic Press.
HAGUE, M.E. (1970), 'A Scale for Describing Young Children', Unpublished dissertation for Dip Ed, University of Birmingham.
HALSEY, A.H. (1972), 'Educational Priority - EPA Patterns and Policies', London, HMSO.
HARRIS, D.B. (1963), 'Children's Drawings as Measures of Intellectual Maturity', New York, Harcourt, Brace & World.

HARRIS, C.C. (1969), 'The Family', London, Allen & Unwin.
HERBERT, G.W. (1972), 'Teachers' Ratings of Classroom Behaviour: A Study of Psychological Correlates among Boys from Working-Class Families', unpublished Ph D thesis, University of Birmingham.
HERBERT, G.W. (1973), Social Behaviour Rating Scale, 'Occasional Papers of the Division of Educational and Child Psychologists', No. 4, 157-63.
HERBERT, G.W. (1974a), Study of preschool age children, 'Child Development Study', Part 3, duplicated report, University of Birmingham.
HERBERT, G.W. (1974b), Teachers' rating of classroom behaviour: factorial structure', 'Brit. J. Educ. Psychol.', 44, 233-40.
HOGGART, R. (1957), 'The Uses of Literacy', London, Chatto & Windus.
HOME OFFICE (1970), 'Report of the Work of the Children's Department 1967-1969, HC 140, London, HMSO.
HOWE, J.R. (1971), 'Two-Parent Families. A Study of their Resources and Needs in 1968, 1969 and 1970', DHSS Statistical Report Series No. 14, London, HMSO.
HUTT, S.J. and HUTT, C. (1970), 'Direct Observation and Measurement of Behaviour', Springfield, Ill., Thomas.
JACKSON, M.S. (1972), 'A Study of the Relationship between Aspects of the Home Background of Infant School Children in Deprived Areas and their Adjustment to School', unpublished M Sc thesis, University College, Swansea.
JENCKS, C. (1972), 'Inequality: A Reassessment of the Effect of Family and Schooling in America', London, Allen Lane, 1973.
JENSEN, A.R. (1969), How much can we boost IQ and scholastic achievement?, 'Harvard Educ. Review', 39, 1-123.
JENSEN, A.R. (1972), 'Genetics and Education', London, Methuen.
JENSEN, A.R. (1973), 'Educability and Group Differences', London, Methuen.
JOSEPH, SIR KEITH (1972), 'The Cycle of Deprivation. 1', Conference of Pre-school Playgroups Association.
JOSEPH, SIR KEITH (1973a), 'The Cycle of Deprivation. 2', Seminar of Association of Directors of Social Services.
JOSEPH, SIR KEITH (1973b), 'The Cycle of Deprivation. 3', National Association for Maternal and Child Welfare.
LEWIS, O. (1967), 'La Vida', London, Secker & Warburg.
LISTER, R. (1974), 'Take-up of Means-tested Benefits', London, Child Poverty Action Group.
LITTLE, A. and MABEY, C. (1972), An index for designation of EPAs, in A. Schonfield and S. Shaw (eds), 'Social Indicators and Social Policy', London, Heineman.
LITTLE, A. and MABEY, C. (1973), Reading attainment and social and ethnic mix of London primary schools, in D. Donnison and D. Eversley (eds), 'London: Urban Patterns, Problems, and Policies', London, Heineman.
LUNZER, E.A. (1955), 'Development of Play Behaviour in Children aged 2-6 Years', unpublished Ph D thesis, University of Birmingham.
LUNZER, E.A. (1970), Construction of a standardised battery of Piagetian tests to assess the development of effective intelligence, 'Research in Education', 3, 53-72.
McALLISTER, J. and MARSHALL, T.F. (1969), The New Junior Maudsley Inventory: Norms for secondary school children aged 11 to 14 years,

'Brit. J. Soc. Clin. Psychol.', 8, 160-3.
McCARTHY, DOROTHY (1954), Language development in children, in L. Carmichael (ed.), 'Manual of Child Psychology', New York, Wiley.
MARSDEN, D. (1968), Autobiographical account, in R. Goldman (ed.), 'Breakthrough', London, Routledge & Kegan Paul.
MIDWINTER, E. (1972), 'Priority Education', Harmondsworth, Penguin.
MILLER, S.M. and ROBY, P. (1970), Poverty: Changing social stratification, in P. Townsend (ed.), 'The Concept of Poverty', London, Heineman.
MITCHELL, S. (1965), 'A Study of the Mental Health of School Children in an English County', unpublished Ph D thesis, University of London.
MORRIS, J.M. (1966), 'Standards and Progress in Reading', Slough, National Foundation for Educational Research.
MULLIGAN, D.G. (1964), 'Some Correlates of Maladjustment in a National Sample of School Children', unpublished Ph D thesis, University of London.
MUSGROVE, F. and TAYLOR, P.H. (1969),'Society and the Teacher's Role', London, Routledge & Kegan Paul.
NATIONAL BOARD FOR PRICES AND INCOMES (1971), 'General Problems of Low Pay', Report No. 169, London, HMSO.
NATIONAL COMMUNITY DEVELOPMENT PROJECT (1974), 'Inter-Project Report', London, CDP Information and Intelligence Unit.
NEWELL, P. (1975), A Free School now, 'New Society', 15 May 1975.
NEWSOM REPORT, see CENTRAL ADVISORY COUNCIL FOR EDUCATION.
NEWSON, J. and NEWSON, E. (1963), 'Infant Care in an Urban Community', London, George Allen & Unwin.
NEWSON, J. and NEWSON, E. (1968), 'Four Years Old in an Urban Community', London, George Allen & Unwin.
PACKMAN, J. (1968), 'Child Care: Needs and Numbers', London, George Allen & Unwin.
PARSONS, T. and BALES, R.F. (1955), 'Family, Socialisation, and Interaction Process', Glencoe, Free Press.
PARSONS, T. (1961), An outline of the social system, in T. Parsons, E. Shils, K.D. Naegele and J.R. Pitts (eds), 'Theories of Society', New York, Free Press.
PATTERSON, G.R., LITTMAN, R.A. and BRICKER, W. (1957), Assertive behaviour in children: a step towards a theory of aggression, 'Monogr. Soc. Res. Child Dev.', 32, 5. Serial no. 113.
PEAKER, G.F. (1967), in Central Advisory Council for Education, 'Children and Their Primary Schools', London, HMSO.
PEAKER, G.F. (1971), 'The Plowden Children Four Years Later', Slough, National Foundation for Educational Research.
PETERS, R.S. (1969), 'Perspectives on Plowden', London, Routledge & Kegan Paul.
PHILLIPS, C.J. and BANNON, W.J. (1968), The Stanford-Binet, Form L-M, Third Revision: a local English study of norms, concurrent validity and social differences, 'Brit. J. Educ. Psychol.', 38, 148-61.
PHILLIPS, C.J., WILSON, HARRIETT, and HERBERT, G.W. (1972), Child Development Study (Birmingham 1968-72), University of Birmingham.
PHILP, A.F. (1963), 'Family Failure', London, Faber & Faber.
PLOWDEN REPORT, see CENTRAL ADVISORY COUNCIL FOR EDUCATION.
POWER, M.L., BENN, P.T. and MORRIS, J.N. (1970), Neighbourhood,

school, and juveniles before the court, 'Brit. J. Criminol.', 12, 111-31.
PRINGLE, M.L., BUTLER, N.R. and DAVIE, R. (1966), '11,000 Seven-Year-Olds', London, Longmans.
RAVEN, J.C. (1956), 'Guide to Using the Coloured Progressive Matrices', London, H.K. Lewis.
RAVEN, J.C. (1958), 'Extended Guide to Using the Mill Hill Vocabulary Scale', London, H.K. Lewis.
REX, J. (1961), 'Key Problems of Sociological Theory', London, Routledge & Kegan Paul.
REX, J. and MOORE, R. (1967), 'Race, Community and Conflict: A Study of Sparkbrook', Oxford University Press.
REYNELL, J. (1969), 'Reynell Developmental Language Scales', Slough, National Foundation for Educational Research.
ROYAL COMMISSION ON THE DISTRIBUTION OF INCOME AND WEALTH (1975), 'Report' No. 1, London, HMSO.
RUNCIMAN, W.G. (1974), Occupational class, and the assessment of economic inequality in Britain, in Dorothy Wedderburn (ed.), 'Poverty, Inequality and Class Structure', Cambridge University Press.
RUTTER, M.L., TIZARD, J. and WHITMORE, K. (1970), 'Education, Health and Behaviour', London, Longmans.
SCHACHT, R. (1971), 'Alienation', London, Allen & Unwin.
SCHAFFER, H.R. and SCHAFFER, E. (1968), 'Child Care and the Family', London, Bell.
SEABROOK, J. (1967), 'The Unprivileged', London, Longmans.
SEARS, R.R., MACCOBY, E.E. and LEVIN, H. (1957), 'Patterns of Child Rearing', New York, Harper & Row.
SHEPERD, M., OPPENHEIM, B. and MITCHELL, S. (1971), 'Childhood Behaviour and Mental Health', University of London Press.
SINFIELD, A. and TWINE, F. (1969), The working poor, 'Poverty', No. 12/13, 4-7.
SMITH, G. and JAMES, T. (1975), The effects of pre-school education: some American and British evidence, 'Oxford Review of Education', 1, 3, 223-39.
SMITH, P.K. and CONNOLLY, K. (1972), Patterns of play and social interaction in preschool children, in N. Blurton-Jones (ed.), 'Ethological Studies of Child Behaviour', Cambridge University Press.
SNEDECOR, G.W. (1956), 'Statistical Methods', Iowa State University Press.
SOUTHGATE, V. (1959), 'Southgate Group Reading Tests - Manual of Instructions, Test 1', University of London Press.
STOTT, D.H. (1966), 'Studies of Troublesome Children', London, Tavistock Publications.
STOTT, D.H. and BALL, R.S. (1965), Infant and Preschool Tests, 'Monogr. Soc. Res. Child Dev.', 30, No. 101.
STUTSMAN, R. (1931), 'Guide for Administering the Merrill Palmer Scale of Mental Tests', New York, World Book Co.
THURSTONE, L.L. and THURSTONE, T. (1953), 'SRA Primary Mental Abilities', USA Science Research Associates.
TIZARD, BARBARA (1974), 'Pre-School Education in Great Britain: A Research Review', London, Social Science Research Council.
TIZARD, J. (1975), Issues in early childhood education, The Dorothy

Gardner Memorial Lecture, 'Child Development Society News Letter', No. 24.
TONGE, W.L., JAMES, D.S. and HILLAM, S. (1975), 'Families Without Hope', London, Royal College of Psychiatrists.
TOWNSEND, P. (ed.) (1970), 'The Concept of Poverty', London, Heineman.
TOWNSEND, P. (1974), Poverty as relative deprivation: resources, and style of living, in Dorothy Wedderburn (ed.) 'Poverty, Inequality, and Class Structure', Cambridge University Press.
TRASLER, G. (1968), Socialisation, in G. Trasler (ed.), 'The Formative Years', London, BBC.
TYERMAN, M.J. (1968), 'Truancy', University of London Press.
VALENTINE, C. (1968), 'Culture and Poverty: Critique and Counterproposals', University of Chicago Press.
VERNON, P.E. (1938), 'The Standardisation of a Graded Word Reading Test', University of London Press.
WALLIS, C.P. and MALIPHANT, R.C. (1967), Delinquent areas in the county of London, 'Brit. J. Crim.', 7, 250-64.
WARD, J. and FITZPATRICK, T.F. (1970), The New British Intelligence Scale: construction of logic items, 'Research in Education', 4, 1-23.
WATTS, A.F. (1958), 'Sentence Reading Test 1', London, National Foundation for Educational Research.
WECHSLER, D. (1949), 'Wechsler Intelligence Scale for Children', New York, The Psychological Corporation.
WEDGE, P. and PROSSER, H. (1973), 'Born to Fail?', London, National Children's Bureau/Arrow Books.
WEST, D.J. (1969), 'Present Conduct and Future Delinquency', London, Heineman.
WEST, D.J. in collaboration with FARRINGTON, D.P. (1973), 'Who Becomes Delinquent?', London, Heineman.
WILKIE, J.S. (1962), 'A Study of the Self Picture and its Concomitants in the Period from the Last Year of the Primary School to the First Year of the Secondary School', unpublished Ph D thesis, University of London.
WILLMOTT, P. (1974), Housing, in M. Young (ed.), 'Poverty Report 1974', London, Temple Smith.
WILSON, HARRIETT (1962), 'Delinquency and Child Neglect', London, Allen & Unwin.
WILSON, HARRIETT (1975), Juvenile delinquency, parental criminality and social handicap, 'Brit. J. Crim.', 15, 241-50.
WINER, B.J. (1962), 'Statistical Principles in Experimental Design', New York, McGraw-Hill.
WOOLF, MYRA (1967), 'The Housing Survey in England and Wales', SS 3 72, London, HMSO.
WRIGHT, C. and LUNN, J.E. (1971), Sheffield Problem Families, 'Community Medicine', 26 November 1971.
WRIGHT, H.F. (1967), 'Recording and Analysing Child Behaviour', New York, Harper & Row.

NAME INDEX

Abel-Smith, B., 32, 233
Askham, J., 34, 233
Atkinson, A.B., 8, 9, 233

Baird, D., 34, 233
Bales, R.F., 88, 237
Ball, R.S., 68, 238
Bannon, W.J., 47, 210, 237
Barnes, J., 12, 180, 189, 190, 233
Bechhofer, F., 182, 235
Becker, H., 181, 233
Benn, P.T., 76, 237
Berger, J., 87, 233
Berman, P.W., 211, 235
Bernstein, B., 193, 233
Birch, H.G., 34, 233
Blurton-Jones, N., 69, 233, 238
Bodmer, W.F., 191, 233
Bonham, D.G., 182, 234
Boxall, M., 6, 233
Brennan, M., 16, 183, 233
Bricker, W., 69, 237
Brimer, M.A., 40, 210, 233
Bruner, J.S., 104, 186, 234
Bullock, A., 11, 234
Burt, C., 191, 234
Butler, N.R., 4, 50, 182, 234, 238

Cane, B., 189, 234
Carmichael, L., 237
Cattell, R.B., 190, 234
Central Advisory Council for Education, 234, 237

Chazan, M., 50, 234
Chief Medical Officer DES, 182, 234
Collison, Lord, 205
Committee on Abuse of Social Security Benefits, 199, 234
Connolly, K., 69, 238
Culyer, A.T., 99, 234

Daniel, W.W., 9, 234
Davie, R., 4, 50, 234, 238
Davies, B., 193, 233
Davies, P., 41, 235
Department of Employment, 23, 27, 28, 200, 203, 234
Department of Health and Social Security, 4, 8, 23, 200, 234
Donnison, D., 235
Douglas, J.W.B., 3, 34, 37, 50, 182, 190, 234
Dunn, L.M., 40, 66, 210, 233, 234
Dunsdon, M.I., 209, 234

Edwards, J., 19, 234
Ernhart, C.B., 211, 235
Evans, R., 41, 235
Eversley, D., 10, 235
Eysenck, H.J., 60, 191, 235
Eysenck, S.B.J., 60, 235

Farrington, P.P., 76, 82, 239
Ferguson, N., 6, 41, 235
Field, F., 8, 199, 233, 234
Fisher Report, 199, 234, 235

242 Name Index

Fitzpatrick, T.F., 212, 239
Furneaux, W.D., 59, 194, 212, 235

Gans, H.J., 7, 235
Gibson, H.B., 50, 59, 60, 194, 212, 213, 217-18, 235
Goldstein, H., 50, 234, 238
Goldthorpe, J.H., 7, 182, 235
Graham, F.K., 211, 235
Greve, J., 10, 235
Greve, S., 10, 235

Hague, M.E., 193, 215, 235
Halsey, A.H., 4, 11, 13, 188, 195, 235
Harris, C.C., 88, 236
Harris, D.B., 210, 235
Herbert, G.W., 46, 49, 50, 55, 59, 70, 82, 192, 214, 236, 237
Hillam, S., 6, 239
Hoggart, R., 7, 88, 98, 120-1, 236
Home Office, 4, 236
Horobin, G., 34, 233
Howe, J.R., 29, 202, 236
Hutt, C., 68, 236
Hutt, S.J., 68, 236

Illsley, R., 34, 233
Inter-Project Report, 11, 237

Jackson, M.S., 50, 234, 236
James, D.S., 6, 239
James, T., 196, 238
Jencks, C., 190, 191, 193, 198, 236
Jensen, A.R., 191, 192, 236
Joseph, K., 186, 236

Kamin, L.J., 191

Lavers, R.T., 99, 234
Levin, H., 105, 106, 133, 175, 238
Leigh, E., 19, 234
Lewis, O., 7, 104, 183-4, 236

Lister, R., 9, 236
Little, A., 11, 12, 189, 236
Littman, R.A., 69, 237
Lockwood, D., 182, 235
Lucas, H., 12, 180, 189, 190, 233
Luckmann, T., 87, 233
Lunn, J.E., 6, 239
Lunzer, E.A., 49, 69, 211, 236

Mabey, C., 11, 12, 189, 236
McAllister, J., 59, 236
McCarthy, D., 74, 237
Maccoby, E.E., 105, 106, 133, 175, 238
Maliphant, R.C., 76, 239
Marsden, D., 87, 237
Marshall, T., 19, 234
Marshall, T.F., 59, 236
Midwinter, E., 13, 237
Miller, S.M., 7, 237
Mitchell, S., 50, 237, 238
Moore, R., 182, 238
Morris, J.M., 188, 237
Morris, J.N., 76, 237
Mulligan, D.G., 50, 237
Musgrove, F., 195, 237

Naegele, K.D., 237
National Board for Prices and Incomes, 28, 237
National Community Development Project, 237
Newell, P., 195, 237
Newsom Report, 11, 234, 237
Newson, E. and J., 3, 16, 89, 105-6, 107-13, 119, 120, 121, 123-5, 127, 130-3, 135-7, 139-42, 145, 147, 149-51, 152, 154, 157, 160, 163, 164-8, 175, 176, 177, 184, 231, 237

Oppenheim, B., 50, 238

Packman, J., 4-5, 237
Page, D., 10, 235
Parsons, T., 6, 88, 181, 237
Patterson, G.R., 69, 237
Peaker, G.F., 189, 190, 237
Peters, R.S., 195, 233, 237

Name Index

Phillips, C.J., 46, 47, 49, 59, 209, 210, 214, 237
Philp, A.F., 34, 237
Pitts, J.R., 237
Platt, J., 182, 235
Plowden Report, 11, 18, 35, 41, 190, 195, 234, 237
Power, M.L., 76, 237
Pringle, M.L., 4, 50, 210, 238
Prosser, H., 4, 8, 34, 182-3, 239

Raven, J.C., 40, 209, 210, 238
Registrar General: Classification of Occupations, 22, 24, 106
Rex, J., 6, 182, 238
Reynell, J., 66, 238
Richardson, K., 233
Richardson, S.A., 34, 233
Roberts, J.A.F., 209, 234
Roby, P., 7, 237
Royal Commission on the Distribution of Income and Wealth, 10, 238
Runciman, W.G., 10, 238
Rutter, M.L., 50, 238

Schacht, R., 88, 238
Schaffer, E. and H.R., 5, 238
Seabrook, J., 7, 238
Sears, R.R., 105, 106, 133, 175, 238
Shaw, S., 234
Shepherd, M., 50, 238
Shils, E., 237
Shonfield, A., 234
Sinfield, A., 26, 238
Smith, G., 196, 238
Smith, P.K., 69, 238
Smithers, J., 189, 234
Snedecor, G.W., 238
Southgate, V., 41, 211, 238
Spears, D., 233

Stott, D.H., 50, 68, 238
Stutsman, R., 66, 238
Supplementary Benefits Commission, 23, 200, 234

Taylor, P.H., 195, 237
Thurstone, L.L. and T., 212, 238
Tizard, B., 196, 238
Tizard, J., 50, 196, 238
Tonge, W.L., 6, 239
Townsend, P., 7, 8, 29, 32, 233, 235, 239
Trasler, G., 105, 239
Twine, F., 26, 238
Tyerman, M.J., 34, 239

Valentine, C., 6, 183, 239
Veit Wilson, J.H., 23-32, 199-208
Vernon, P.E., 41, 211, 239

Wallis, C.P., 76, 239
Ward, J., 212, 239
Watts, A.F., 41, 211, 239
Wechsler, D., 40, 209, 210, 239
Wedderburn, D., 233, 235, 238, 239
Wedge, P., 4, 8, 34, 182-3, 239
West, D.J., 4, 34, 50, 54, 76, 82, 192, 217, 239
Whitmore, K., 50, 238
Wilkie, J.S., 59, 239
Williams, A., 99, 234
Williams, P., 6, 41, 235
Willmott, P., 10, 239
Wilson, H., 34, 46, 49, 59, 76, 77, 79, 80, 214, 237, 239
Winer, B.J., 48, 239
Woolf, M., 20, 239
Wright, C., 6, 239
Wright, H.F., 69, 239

SUBJECT INDEX

ability, see tests of ability and attainment
anxiety, 51, 55-6, 58, 193, 214
see also adjustment problems, behaviour ratings
accidents to children, 96
activities
of 3-4 year olds, 108-10, 123-5, 144, 169
of 6-7 year olds, 110-13, 122, 125-7, 135-8, 144-5, 170
of 10-11 year olds, 113-16, 122, 128-9, 138-9, 144-5, 170
see also family activities
'adaptational approach', 184-6
adjustment problems, 5, 12-14, 50, 55-9, 63, 82-3, 181, 192-4
see also behaviour ratings, behavioural problems
aggression
control of, 139-44
in school, 51, 53, 55-9, 61
alienation, 88
attainments
and family characteristics, 170, 191-2
and social characteristics of area, 11-12, 188-90
and social handicap, 33-5
of school-age boy, 41-9, 62, 187-90
see also tests of ability and attainment
attention-seeking, 51, 53, 55-6, 72, 192
authoritarian-democratic dimension, 107, 164

bedtime arrangements, 124-5, 129, 130-1
behaviour ratings
correlations with self report, 61-2, 222
description of scales used, 50-1, 214-18
results, 51-9, 62-3, 192-4, 220-2
review, 12-13, 50
behavioural problems in school, 12-14, 82-3, 170, 178-9
see also adjustment problems
Bristol Social Adjustment Guides, 50

chaperonage, 145-6, 168-79, 185
checklist of behaviour
see play groups
'child-centredness', 106, 107, 150, 164-6, 168-77
children's department
see social services department
children's jobs around the house, 123-4, 135-9, 144-5, 149-50, 170, 176
'collective orientation', 186
competence, 13, 53-7, 62, 185, 193, 196
see also behaviour ratings, motivation
community development projects, 10-11
community schools, 195
conduct scale, 50, 54-7, 217, 220, 222
confinement, 96-7

244

contraception, 21, 37
copying designs, 49, 68, 211-12, 219
'cultural deprivation', 6
culture clash, 13, 195
'culture of poverty', 7, 104, 183-4

delinquency, 17-18, 76-83, 113-15, 170-4, 179
deprivation
  areas of, 10-12, 18-19, 89-90, 180, 182, 195
  definitions, 6-7, 35
  see also disadvantaged children
disadvantaged children, 4-6, 8, 11-12, 14, 190, 191, 196-8
  see also social handicap
Draw-a-Man test, 43, 45, 48, 210, 213-14, 219, 220, 221, 222
earnings
  levels in population, 8-9, 26-9, 181-2
  sample families, 23-32, 199-208
  see also income distribution among sample
education
  'compensatory', 6, 14, 41
  new concepts of, 195
  parental, 22
  parental attitudes towards, 155-60, 186
educational priority areas, 11-12, 41, 188-90, 195
English Picture Vocabulary Test, 40, 42, 44, 47, 210, 219, 221, 222
environment
  hazards of as perceived by parents, 123, 126-7, 128-9, 145-6, 172, 177
  of sample families, 18-19

family activities, 90-2, 116-22, 169, 170
Family Service Unit, 17, 113
family size
  and behaviour problems, 50
  and composition of sample, 18, 21-2, 96, 99
  and deprived areas, 36-7
  and income, 29, 203
  and performance at school, 34
  and social class, 3, 106, 184-5
friendships
  children's, 108-9, 110-11, 113-14
  mothers', 96-7, 103

general practitioners, 93-6
genetic aspects, 6, 33, 191-2, 198
Gibson Spiral Maze, 213

health
  and poverty, 182-3
  of sample families, 92-3, 99-100, 183
  see also illness, impairment
health visitors, 95-6
heritability, 191-2
  see also genetic aspects
home atmosphere, 100-2, 121-2, 170-2
housing
  costs, 32, 203-4
  effects on child development of inadequate, 5, 34, 76, 106-7, 164, 184-5
  parental comments, 89-90
  and the poor, 10-11, 182
  of sample families, 19-21

ILEA Literacy Survey, 11-12, 189
illness
  attitudes of parents towards, 93
  incidence, 92-3, 183
  strategies to combat, 93-8
  see also health, impairment
immediate memory, 49, 212
impairment
  of children, 93, 99, 183
  of parents, 92, 99, 183
income distribution among sample, 18, 23-32, 199-208
independence training, 123-9, 144
individual components of variation, 49
induction test, 212, 219, 221
intergroup differences

see behaviour ratings, tests of ability and attainment, self-report scales, play groups
intervention
  nursery and infant schools, 14, 196-7
  at later school ages, 189-90, 197-8
introversion-extraversion, 51, 61, 194-5, 215
  see also self-report scales
language
  development (play groups), 66, 68, 72, 74
  as used in directing behaviour, 152-5, 160, 166-8, 185
  see also play groups, observations of language
linearity of trend, 48-9, 54-5

manipulation, 66, 68, 71
marital relations, 100-1
mental retardation, 5, 6, 92-3
Merrill-Palmer Scale, 66-8
Mill Hill Vocabulary Test, 40-2, 209, 219, 221
mobility
  geographical, 10, 20-1, 88
  social, 7, 22, 88, 181
modesty training, 129-35, 145, 170, 176
mothers' ability to cope, 91-2, 100-2
motivation
  behaviour rating scales, 50, 55
  self-report scales, 59-62
  see also competence
motor dots test, 49, 212, 219, 221

narrative recording, 69-70
National Child Development Study, 3-4, 8, 50, 182-3, 210
National Society for the Prevention of Cruelty to Children, 17
neighbours, 88, 89-90, 96, 97, 140-2, 163, 184
New Junior Maudsley Inventory, 59-61, 194, 212, 222

NFER Reading Test A-D, 41, 46-7, 55, 211, 220, 222
nursery schools, 196-7

occupations of parents, 22

parental interests and hobbies, 117-19, 170
parental participation in children's activities, 116-22, 168-72
'pathology model', 5, 6, 7, 181, 183
Peabody Picture Vocabulary Test, 66-7
pencil, use of, 44, 49, 68, 212
permissive-restrictive dimension, 137, 147-55, 175-7, 185
Piagetian tests (Lunzer), 49, 211, 219, 221
play, 108-13, 115, 163, 185
play groups
  observations of behaviour, 68-72
  observations of language, 72-5
  psychological tests, 65-8
play, maturity of
  see play groups, observation of behaviour
pocket money, 111-12, 115-16, 163
political power, 7, 181
possessions of child, 109, 112, 185
  see also toys
poverty
  definitions, 7-8, 29, 164, 202-3
  effects on child development, 5, 34, 126, 144, 163-4, 198
  of sample, 18, 26-32, 98-9, 164
  studies, 7-11
  syndrome, 181-3
  trap, 9, 185-6
powerlessness, 7, 181, 185
punishment
  smacking, 147-52, 160, 167
  other, 152-55, 167

quarrels, mothers' intervention, 140-4

Raven's Coloured Progressive
  Matrices, 40-2, 45, 48, 210,
  219, 220, 221, 222
reading, 41, 46-9, 55, 62, 187-
  90, 197, 210-11
relatives, 88, 96, 97, 102-3
research methods and objectives,
  15-17, 33-6, 77, 87-9, 105-8,
  168-9, 180-1, 183-4, 199-204,
  209-18, 223-31
Reynell Developmental Language
  Scales, 66-7
rules of behaviour made explicit,
  126-7, 129, 145, 175-7

sampling bias in control groups,
  46-7, 49
savings, 9-10, 112, 181
school
  attendance, 11, 34-5, 156-7,
    158, 160, 170, 194, 197, 198
  children's complaints, 156-60
  parental contact, 35, 39, 155
  parental help with school work,
    119-20
Schools Council Project on compensatory education, 41
self-concept
  see self-report scales
self-report scales
  correlation with reading tests,
    62, 195, 222
  description, 59, 212-13
  results, 60-3, 194-5, 222
  see also behaviour ratings
sentence repetition test, 49, 212,
  219, 221
'situational approach', 7, 184
social class
  and child-rearing practices,
    105, 106, 107, 130, 135, 147,
    152, 163, 164, 166, 167, 184-6
  distribution in deprived areas,
    36-7
social deprivation
  see deprivation
social handicap, 18, 33-9, 180,
  187-8, 192-5
  see also intergroup differences
social interaction
  see play groups, observations
  of behaviour
socialization, 6, 87-9, 105-7

social mobility
  see mobility
social network, 103
social services department, 3, 4,
  15, 17-18, 36, 97, 176, 180-1,
  183, 196
sociological models
  conflict, 6-7, 181
  structural-functional, 6, 7,
    88, 181
Southgate Reading Test 1, 41-2,
  55, 211, 219, 221
streaming, 46
stress, 3, 4-5, 98-102, 104, 164,
  181, 183, 196
strictness
  see permissive-restrictive
  dimension
sub-cultural continuities, 6-7,
  184-6
subnormality
  see mental retardation
supervision of children
  see chaperonage
supplementary benefit, 23-32,
  199-208
  see also income distribution
  among sample

teachers, 12-14, 156-60, 188-9,
  190, 193-4, 195, 196-7
  see also behaviour ratings
tests of ability and attainment
  description, 40, 209-14
  and disadvantaged children,
    190
  results, 41-9, 62-3, 187-8
  see also play groups
threats by mothers, 153-5, 160
time-sampling, 68-70
toys, 109-10, 112, 113, 170
  see also possessions of child
truancy
  see school, attendance
truth, evasion or distortion of,
  106, 166-8

Vernon Graded Word Reading Test,
  41, 45, 47-8, 211, 220, 222

welfare clinics, 95-6

Wechsler Intelligence Scale for
  Children
  block design subtest, 40-1,
    44, 48, 210, 220, 222
  coding subtest, 40-2, 44, 48,
    210, 220, 222
  vocabulary subtest, 40-2, 48,
    209, 220, 222
White Lion School, 195
work record of fathers, 18, 23-6,
  200

For Product Safety Concerns and Information please contact our
EU representative GPSR@taylorandfrancis.com Taylor & Francis
Verlag GmbH, Kaufingerstraße 24, 80331 München, Germany